LIE GROUPS

FOR

PEDESTRIANS

by

HARRY J. LIPKIN

The Weizmann Institute of Science
Rehovot, Israel

DOVER PUBLICATIONS, INC.
Mineola, New York

Bibliographical Note

This Dover edition, first published in 2002, is an unabridged republication of the second edition (1966) of the work originally published by North-Holland Publishing Company, Amsterdam, in 1965.

Library of Congress Cataloging-in-Publication Data

Lipkin, Harry J.
Lie groups for pedestrians / Harry J. Lipkin.
p. cm.
"This Dover edition . . . is an unabridged republication of the second edition (1966) of the work originally published by North-Holland Publishing Company, Amsterdam, in 1965"—T.p. verso.
Includes bibliographical references and index.
ISBN 0-486-42185-6 (pbk.)
1. Particles (Nuclear physics) 2. Lie groups. I. Title.

QC782 .L56 2002
539.7'2'0151255—dc21

2002022117

Manufactured in the United States of America
Dover Publications, Inc., 31 East 2nd Street, Mineola, N.Y. 11501

PREFACE TO FIRST EDITION

As a graduate student in experimental physics, I found the study of group theory considered to be a useless 'high-brow' luxury. Furthermore all attempts to follow a lecture course resulted in a losing battle with characters, cosets, classes, invariant subgroups, normal divisors and assorted lemmas. It was impossible to learn all the definitions of new terms defined in one lecture and remember them until the next lecture. The result was complete chaos.

It was a great surprise to find later on that (1) techniques based on group theory can be useful; (2) they can be learned and used without memorizing the large number of definitions and lemmas which frighten the uninitiated. Angular momentum is presented in elementary quantum mechanics courses without a detailed analysis of the Lie group of continuous rotations in three dimensions. The student learns about angular momentum multiplets and coupling of angular momenta without realizing that these are the irreducible representations of the rotation group. He also does not realize that the algebraic properties of other Lie groups can be applied to physical problems in the same way as he has used angular momentum algebra, with no need for characters, classes, cosets, etc.

This book began as a short article with the aim of presenting the 'group theoretical' methods used in nuclear structure in a simple way. Another short article was begun to point out that bilinear products of creation and annihilation operators lead to Lie algebras, and to classify the algebras obtained in this way. These were then combined with a discussion of 'quasispin' operators acting like fictitious angular momenta which arise in various areas in physics. This material, now in Chapters 4 and 5, was presented in a series of lectures at Argonne National Laboratory in the summer of 1961, discussing simple models of many-particle systems and the application of group theory. The article thus became a set of lecture notes.

The Argonne lecture notes were still unfinished when unitary symmetry appeared and created a demand from high energy physicists for intelligible lectures on group theory. They wanted to understand and use unitary symmetry without learning about characters and cosets. A series of lectures was given at the University of Illinois and the lecture notes had a different emphasis from the Argonne notes. The audience was interested in unitary symmetry and elementary particles, not in nuclear structure and many-body problems. After several revisions and additions the lecture notes from Illinois and Argonne were combined and extended to form this book.

The aim of the book is to show how the well-known techniques of angular momentum algebra can be extended to treat other Lie groups, and to give several examples illustrating the application of the method. Because of the present interest in symmetries of elementary particles, this particular application is stressed. Chapter 1 presents the essential features of the method by analogy with angular momentum and points out that bilinear products of creation and annihilation operators lead to Lie algebras. Chapter 2 presents isospin as the first example of the method. Chapter 3 presents the group SU_3 and its application to elementary particles. Chapter 4 gives the treatment of the three-dimensional harmonic oscillator using SU_3 and discusses its application to nuclear structure. Chapter 5 considers the classification of Lie algebras of bilinear products of creation and annihilation operators, symplectic groups, and the applications to pairing correlations and seniority in many-particle systems. Chapter 6 discusses permutation symmetry and gives a simplified version of Young diagrams as a guide to their use.

The appendices constitute a large portion of the book and present a detailed study of the application of SU_3 algebra to unitary symmetry of elementary particles. Appendix A builds up the structure of the SU_3 multiplets by combining fundamental triplets. Appendix B develops the U-spin method for calculating experimental predictions from unitary symmetry. Appendix C presents many detailed examples of experimental predictions from unitary symmetry. Appendix D is a short discussion on the phases which plague all investigators.

I should like to express my appreciation to many colleagues at the University of Illinois and Argonne National Laboratory who forced me to explain this material to them in a series of constantly interrupted lectures, and to the secretarial staff, particularly M. Runkel and E. Kinstle who performed the incredible job of getting the notes out almost before the lectures were given. It is a pleasure to thank Y. Ne'eman for introducing me to unitary symmetry and C. A. Levinson and S. Meshkov for showing me how the techniques they developed for nuclear structure could be used for elementary particles. I should also like to thank G. Racah for many stimulating discussions and to acknowledge having learned a great deal from a series of his seminar lectures which showed how many useful results could be obtained with the use of simple but powerful algebraic methods. Finally I should like to thank all my colleagues at the Weizmann Institute who helped in the preparation of this book, and particularly L. Mirvish, who typed the manuscript, R. Cohen, who prepared the figures, and H. Harari for criticism of the manuscript.

PREFACE TO SECOND EDITION

At the time when the first edition of this book was going to press new symmetries of elementary particles appeared which were based on the Lie groups SU_4, SU_6 and SU_{12}. These groups were not discussed in any detail in the first edition. However, the general approach is easily extended to treat such groups of rank three and higher. Thus at the time of a second printing of this book, it appears that the addition of material on these groups would be useful to the reader, even though only a short time has elapsed since the first edition went to press.

This second edition was prepared with the aim of incorporating useful material on groups of higher rank with a minimum of modification of the book in order to allow it to appear as soon as possible. Chapter 7 has been added to present material on the groups SU_4, SU_6 and SU_{12}. This additional chapter was written in a style as close as possible to that of the preceding work. There has been no revision of the previous text except for the correction of a few minor errors.

It is dangerous to attempt to keep a book like this exactly up to date with current research developments. However, it seems as if groups of rank higher than two should be of interest to physicists for some time in the future. The purpose of Chapter 7 is not to present an up-to-date picture of the present status of elementary particle symmetries, but rather to show how groups of higher rank can be treated with the same 'pedestrian' approach presented earlier for groups of rank two. The material on elementary particles may be only of historical interest by the time the book appears.

However, the general algebraic techniques should still be useful. For this reason no attempt has been made to include detailed comparisons of higher symmetry predictions with experimental data. The Sakata model has been retained for the fundamental triplet of SU_3, to keep the same style as in the earlier chapters without rewriting them.

Another development since the appearance of the first edition of this book has been the sudden interest in non-compact groups and their infinite-dimensional representations. The possibility of expanding the very brief treatment of this subject in § 5.6 was considered, but is not feasible at this time.

At the time of writing of this preface the book 'High Energy Physics and Elementary Particles' published by the International Atomic Energy Agency, Vienna 1965, has recently appeared. The reader is referred to this book for more detail comparisons of higher symmetries with experimental data as well as discussions of mathematical theory of non-compact algebras and their applications to elementary particles. However, the characteristic difficulty in the preparation of such material is also evident in this book despite its rapid preparation. New experimental data have already made some of the material obsolete.

I should like to express my appreciation to H. Harari whose Ph.D. thesis (Hebrew University, Jerusalem – in Hebrew) contains many useful tables which were an aid in the preparation of Chapter 7, and to D. Agassi and C. Robinson for pointing out errors in the first edition.

CONTENTS

Chapter 4

THE THREE-DIMENSIONAL HARMONIC OSCILLATOR

Chapter 5

ALGEBRAS OF OPERATORS WHICH CHANGE THE NUMBER OF PARTICLES

Chapter 6

PERMUTATIONS, BOOKKEEPING AND YOUNG DIAGRAMS

Chapter 7

THE GROUPS SU_4, SU_6 AND SU_{12}, AN INTRODUCTION TO GROUPS OF HIGHER RANK

Appendices

CHAPTER 1

INTRODUCTION

Physicists have not yet learned to live with group theory in the same way as they have learned for other mathematical techniques such as differential equations. When an experimentalist or advanced graduate student encounters a simple differential equation in the course of his work, he does not run away and hide, worry about whether the solution to the equation really exists, or indulge in mathematical exercises of a 'high-brow' nature. He either solves the equation or goes to the literature and looks up the solution. On the other hand, many sophisticated theorists who are quite at home in the complex plane seem to be afraid of what might be called elementary exercises in group theory. This is all the more mysterious since many of these so-called group theoretical methods are in principle no different and no more complicated than certain mathematical techniques which every physicist learns in a course in elementary quantum mechanics; namely, the algebra of angular momentum operators.

The reason for this difficulty may be that physicists have still not made the separation analogous to that made for differential equations between those parts of the subject which belong to the physicist and those which belong to the mathematician. The standard treatment of group theory for physicists begins with complicated definitions, lemmas, and existence proofs which are certainly necessary for a proper understanding of group theory. However, it is possible for physicists to understand and to use many techniques which have a group theoretical basis without necessarily understanding all of group theory, in the same way as he now uses angular momentum algebra without delving deeply into the mysteries of the three-dimensional rotation group.

The purpose of this treatment is to show how techniques analogous to angular momentum algebra can be extended and applied to other group theoretical problems without requiring a detailed understanding of group theory.

1.1. REVIEW OF ANGULAR MOMENTUM ALGEBRA

Consider three angular momentum operators J_x, J_y and J_z which satisfy the well-known commutation rules

$$[J_x, J_y] = \mathrm{i}J_z, \qquad [J_y, J_z] = \mathrm{i}J_x, \qquad [J_z, J_x] = \mathrm{i}J_y. \tag{1.1}$$

From these commutation rules it follows that there exists an operator

$$J^2 = J_x^2 + J_y^2 + J_z^2$$

which has the property of commuting with all the angular momentum operators:

$$[J^2, J_x] = [J^2, J_y] = [J^2, J_z] = 0. \tag{1.2}$$

Since J^2 commutes with all the operators, it commutes with any one of them, and one usually chooses the operator J_z. One can then in any problem find a complete set of states which are simultaneous eigenfunctions of J^2 and J_z with eigenvalues usually designated by J and M. We use the conventional designation for these states

$$|J, M\rangle. \tag{1.3}$$

The remaining two operators J_x and J_y do not commute with J_z, but the following simple linear combinations

$$J_+ = (J_x + \mathrm{i}J_y) \quad \text{and} \quad J_- = (J_x - \mathrm{i}J_y) \tag{1.4}$$

satisfy particularly simple commutation rules. Since J^2 commutes with all the operators, we have

$$[J^2, (J_x + \mathrm{i}J_y)] = [J^2, (J_x - \mathrm{i}J_y)] = 0. \tag{1.5}$$

The commutators with J_z are also quite simple,

$$\begin{aligned} [J_z, (J_x + \mathrm{i}J_y)] &= (J_x + \mathrm{i}J_y), \\ [J_z, (J_x - \mathrm{i}J_y)] &= -(J_x - \mathrm{i}J_y). \end{aligned} \tag{1.6}$$

The commutator of each of these operators with J_z is just the same operator again, multiplied by a constant. It then follows that if either of these operators operates on a state which is a simultaneous eigenfunction of J^2 and J_z with eigenvalues J and M, the result is another state which is an eigenfunction of J^2 with the same eigenvalue J and which is also an eigenfunction of J_z, but with the eigenvalue $M \pm 1$.

$$(J_x \pm iJ_y)|J, M\rangle = \sqrt{J(J+1)-M(M\pm 1)}\,|J, M\pm 1\rangle . \quad (1.7)$$

The value of the coefficient appearing on the right-hand side is easily obtained by a little algebra. This result and the trivial

$$J_z|J, M\rangle = M|J, M\rangle \quad (1.8)$$

define matrix elements for all of the angular momentum operators for all of the complete set of states.

Beginning with any particular state, $|J, M\rangle$, a set of states can be generated by operating successively with the operators $(J_x + iJ_y)$ and $(J_x - iJ_y)$. This process cannot be continued indefinitely because M can never be greater than J. Thus one finds restrictions on the possible eigenvalues of J and M, and obtains the well-known result that these may be either integral or half-integral and that for any eigenvalue J there corresponds a set or multiplet of $2J+1$ states all having the same eigenvalue of J and having values of M equal to $-J$, $-J+1$, ..., $+J$. The full set of states in a multiplet can be generated from any one of the states by successive operation with the operators $(J_x \pm iJ_y)$.

Some of these features can be demonstrated simply in diagrams of the type shown in Fig. 1.1. These diagrams are one-dimensional plots of the eigenvalues of J_z. Fig. 1.1a represents the operators $(J_x + iJ_y)$ and $(J_x - iJ_y)$ as vectors which change the eigenvalue of J_z by ± 1, respectively. Figure 1.1b illustrates the structure of a typical multiplet, in this case one with $J = \frac{7}{2}$, in which a point is plotted for each value of J_z where a state exists in the multiplet. The operation of any of the operators in Fig. 1.1a on the states in the multiplet of Fig. 1.1b is represented graphically by taking the appropriate vector of Fig. 1.1a, placing it on Fig. 1.1b and noting which states are connected by this vector.

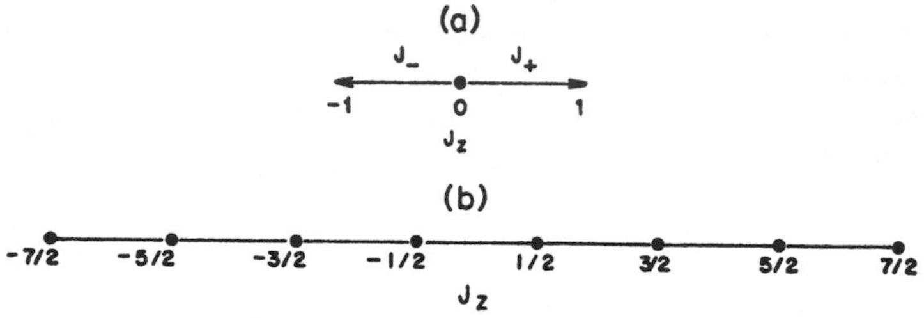

Fig. 1.1

There are also well-known rules for combining multiplets. A system may consist of several parts, each of which is characterized by a multiplet having a particular value of J. (An example of this would be the orbital and spin angular momenta of a particle.) These multiplets can be combined to form a multiplet describing the whole system. (For example, an orbital angular momentum of 2 and a spin of $\frac{1}{2}$ for a particle can be combined to give a total angular momentum either of $\frac{3}{2}$ or $\frac{5}{2}$.) Given the J-values for the multiplets describing parts of the system, there are simple rules for deciding which possible values of J occur for the total system, and there are algebraic techniques involving vector coupling coefficients for expressing the wave functions of the combined system which belong to a given multiplet. There is also one very simple rule which results from the different character of the multiplets having half-integral and integral values of J. If two multiplets having integral values of J are combined, the multiplet describing the overall system must also have integral values of J. If two multiplets having half-integral values of J are combined the multiplets describing the combined system must also have integral values of J. On the other hand, if a multiplet having an integral value of J is combined with one having a half-integral value of J, the multiplet describing the combined system then must have a half-integral value of J.

So far we have considered only the consequences of the angular momentum commutation rules (1.1) and have made no mention of any Lie group. All the results obtained thus far therefore depend only on the existence of operators satisfying the angular momentum commutation rules and are not in any way dependent upon the existence of a continuous group of transformations. Let us now consider briefly the relation between the angular momentum operators and the group of continuous rotations in three dimensions. It is well known that the operators J_x, J_y and J_z can be considered as generators of infinitesimal rotations. For example, if ψ is a wave function for a particular system and J_x is a total angular momentum operator for that system then the wave function

$$\psi' = (1 + i\varepsilon J_x)\psi \tag{1.9}$$

represents the state ψ rotated by infinitesimal angle ε about the x-axis. Similar relations exist for infinitesimal rotations about the y- and z-axes and finite continuous rotations can be generated from these infinitesimal rotations. The statement that a Hamiltonian is invariant under rotations is equivalent to the statement that it commutes with the three angular momentum operators, since the latter generate infinitesimal rotations from which all the finite rotations can be built. By studying the properties of these rotations and the way wave functions and operators transform under them, many interesting and useful results can be obtained. These, however, are not considered in this treatment. Rather we confine ourselves to those results which are obtainable simply from the algebra of the angular momentum operators; i.e. from the properties of the generators of the infinitesimal rotations.

The algebraic relations among the angular momentum operators are useful in physical problems because these operators are often simply related to other operators which describe properties of a physical system. Examples of such operators are the Hamiltonian, electric and magnetic moments, and operators inducing transitions. Operators of physical interest often satisfy very simple commutation relations with the angular momentum operators. This is of course equivalent to stating that the corresponding physical quantities transform in a simple way under rotations. The simple transformation properties are expressed mathematically by the classification of operators as scalars, vectors, tensors, etc. These simple transformation properties are also expressible as simple commutation relations. A scalar operator which is invariant under rotations commutes with all the angular momentum operators. A vector operator consists of three components which transform into combinations of one another under rotations and which satisfy commutation relations with the angular momentum operators analogous to those of the angular momentum operators among themselves.

$$[J_x, S] = [J_y, S] = [J_z, S] = 0 \tag{1.10a}$$

$$[J_x, V_y] = \mathrm{i}V_z, \qquad [J_y, V_z] = \mathrm{i}V_x, \qquad [J_z, V_x] = \mathrm{i}V_y \tag{1.10b}$$

where S is a scalar operator and (V_x, V_y, V_z) are the components of a vector operator.

More generally, operator multiplets can be defined in a manner analogous to the wave function multiplets. These operator multiplets are called irreducible tensors and their components transform into linear combinations of one another under rotations. The commutators of such tensor operators with the angular momentum operators have the same form as the step operator relations (1.7) and (1.8) for angular momentum multiplets

$$[(J_x \pm \mathrm{i}J_y), T_{kq}] = \sqrt{k(k+1) - q(q \pm 1)}\, T_{k(q\pm 1)} \tag{1.11a}$$

$$[J_z, T_{kq}] = qT_{kq} \tag{1.11b}$$

where T_{kq} is the q-component of an irreducible tensor of degree k, and the indices k and q are analogous to the angular momentum eigenvalues J and M for the corresponding wave function multiplet. Such an irreducible tensor has $2k+1$ components and the index q takes on $2k+1$ values from $-k$ to $+k$.

From the commutation relations (1.11) it follows that irreducible tensors combine in the same way as angular momentum multiplets and that the calculation of matrix elements of irreducible tensor operators between angular momentum eigenstates satisfies angular momentum coupling rules. This is expressed quantitatively by the Wigner–Eckart theorem which states that the matrix elements of different components of the same irreducible tensor between states which are members of the same two angular momentum multiplets are all proportional to one another with coefficients depending only upon angular momentum algebra and independent of the intrinsic properties of the operators. All these matrix elements are thus determined by a single number characteristic of this operator: the so-called 'reduced matrix element'.

Let us now review in more detail how angular momentum algebra is used in physical problems. We consider first the application of angular momentum algebra to the solution of the time-independent Schrödinger equation. There are several possibilities depending on the form of the commutation relation between the Hamiltonian and the angular momentum operators.

(1) The Hamiltonian commutes with all the angular momentum operators. Then a complete set of eigenstates of the Hamiltonian can be found which are also eigenfunctions of J^2 and J_z and the $2J+1$ states of a multiplet are all degenerate eigenstates of the Hamiltonian. The use of the angular momentum algebra therefore simplifies the solution of the eigenvalue problem for the Hamiltonian by defining two integrals of the motion; i.e. two quantum numbers which can be used to specify the eigenstates of the Hamiltonian. Note that J_y or J_x could also be chosen instead of J_z to specify the states, which would then be linear combinations of the eigenfunctions of J_z.

(2) The Hamiltonian does not commute with all the angular momentum operators but still commutes with J^2 and J_z. An example of this case is the motion of a spinless charged particle in a spherically symmetric field with an additional uniform magnetic field in the z-direction. The Hamiltonian for this case would have the form

$$H = H_0 + KJ_z \tag{1.12}$$

where H_0 commutes with all the angular momentum operators and the K is a constant. For this case, the eigenstates of the Hamiltonian can still be chosen to be simultaneous eigenfunctions of J^2 and J_z, but the $2J+1$ states of a multiplet are no longer degenerate. The splitting of the energy levels in a multiplet is determined by the terms in the Hamiltonian that do not commute with J_x and J_y.

For the example (1.12) the energy levels within the multiplet are equally spaced with a splitting K between adjacent levels, as a result of the term KJ_z. Note that for this case it is not possible to choose J_x or J_y rather than J_z to specify the states, as J_x and J_y do not commute with the Hamiltonian.

(3) The Hamiltonian does not commute with all the angular momentum operators, but the commutators have a simple form. The example (1.12) satisfies this criterion, since

$$[H, (J_x \pm iJ_y)] = \pm K(J_x \pm iJ_y) . \tag{1.13}$$

The commutators can be used to determine properties of the energy spectrum of the Hamiltonian. For example, given any eigenfunction ψ of the Hamiltonian (1.12) we can generate other eigenfunctions using the commutator (1.13).

$$H\psi = E\psi , \tag{1.14a}$$

$$H(J_x \pm iJ_y)\psi = (E \pm K)(J_x \pm iJ_y)\psi . \tag{1.14b}$$

This result (1.14) leads again to the conclusion that the energy spectrum of the Hamiltonian (1.12) consists of sets of equally spaced energy levels with a spacing K between adjacent members. Although in this simple case these conclusions were evident by inspection of the Hamiltonian (1.12), there are many other cases where commutation relations analogous to (1.13) lead to non-trivial results.

The commutation relations (1.13) can be considered as the equations of motion of the operators $J_x \pm iJ_y$. The content of equations (1.13) and (1.14) can also be described by saying that operators which 'satisfy simple equations of motion' can be used to generate excitations of a system.

(4) Any of the properties above apply not to the exact Hamiltonian but to an approximate Hamiltonian which is used as a basis of perturbation theory. The treatment of the unperturbed Hamiltonian is then simplified by the use of the angular momentum algebra as described above.

If the perturbation is expressed simply in terms of irreducible tensors, the first-order perturbation result for the energy splitting within a given angular momentum multiplet is given directly by the Wigner–Eckart theorem in terms of a single parameter depending on the perturbation. Common examples are the Zeeman and quadrupole splittings arising when nuclear, atomic or molecular systems are placed in external fields.

The angular momentum algebra is used in a similar manner in time-dependent problems such as decay and reaction processes where one considers a transition between an initial state and a final state. If the Hamiltonian commutes with all the angular momentum operators, then angular momentum must be conserved in the transition from the initial state to the final state. If the initial state is not an angular momentum eigenfunction (e.g. a plane wave) it is often useful to expand it in angular momentum eigenfunctions (partial wave expansion) because of the conservation of angular momentum. Each angular momentum eigenvalue (partial wave) then defines an independent channel for the reaction which is uncoupled from the other channels. This decoupling of the different channels allows the scattering process to be treated separately for each channel, thereby greatly simplifying the solution of the Schrödinger equation.

1.2. GENERALIZATION BY ANALOGY OF THE ANGULAR MOMENTUM RESULTS

Starting from the simple commutation rules (1.1) we have found a classification scheme for the states of a quantum mechanical system in which angular momentum operators have simple properties. The states are divided into sets or multiplets such that the matrix elements of all angular momentum operators vanish between states belonging to different multiplets. Within each multiplet the action of the angular momentum operators is very simple. Appropriate linear combinations of these operators can be chosen such that they are either diagonal like J_z or 'step operators' like $J_x \pm iJ_y$. The latter simply change the eigenvalue of the state on which they are operating and thus jump from one state to another through the multiplet.

We now assert without proof: *Whenever one encounters a set of operators satisfying similar commutation rules, one can play the same game*. One can define multiplets and suitable linear combinations of the operators such that these operators are either diagonal in the representation defining the multiplets or act like step operators within a multiplet. The matrix elements of all operators between states of different multiplets vanish. Those readers who are interested in the general proofs underlying these assertions are referred to the standard literature on group theory. Those who are not interested in the proofs can see in the material which follows examples of how the game can be played quite usefully in specific cases by simply going ahead and constructing the representations and the multiplets.

In the remainder of this section we state more precisely what exactly is meant by 'playing the game'. In the following section some general arguments concerning the algebra of second quantized operators are given to show why such sets of operators can be expected to occur frequently in physics. In the remainder of this book we deal with a number of specific examples showing how the technique can be used.

Let us now assume that we have a finite number of operators X_ϱ which satisfy commutation rules similar to those of the angular momentum operators; namely that the commutator of any two of the operators is a linear combination of the operators of the set:

$$[X_\varrho, X_\sigma] = \sum C^\tau_{\varrho\sigma} X_\tau \tag{1.15}$$

where the coefficients $C^\tau_{\varrho\sigma}$ are constants. A set of operators satisfying such commutation rules is called a Lie algebra. We now assert that from these operators we can construct operators like J^2 which commute with all the operators of the set. There may be only one such independent operator, like J^2 in the case of angular momenta, or there may be more than one. We shall call such operators C_μ. These operators are sometimes called Casimir operators.

$$[C_\mu, X_\varrho] = 0 \quad \text{for all} \quad \mu, \varrho\,. \tag{1.16}$$

We now choose one of the operators of the set, as we chose J_z for angular momentum, to be diagonal in the representation we shall define. It may be possible to choose more than one such operator.

If there are many operators in the set there may be operators within the set which commute with one another. We shall choose as many commuting operators as we can find and denote them by the letters H_i. These operators H_i also commute with the Casimir operators C_μ, since the latter commute with all the operators of the set. We can therefore find a complete set of states in any problem which are simultaneous eigenfunctions of all the operators C_μ and H_i, with eigenvalues C'_μ and H'_i

$$|C'_\mu, H'_i\rangle\,, \tag{1.17}$$

analogous to the complete set of states $|J, M\rangle$ for angular momentum. We now further assert that the remaining operators of the set can all be expressed in terms of a linearly independent set of step operators. We call these operators

$$E_\alpha \tag{1.18}$$

and state that they satisfy the following simple commutation rules:

$$[C_\mu, E_\alpha] = 0\,, \tag{1.19}$$

$$[H_i, E_\alpha] = \alpha_i E_\alpha\,. \tag{1.20}$$

These are directly analogous to the commutation rules satisfied by the operators $J_x \pm iJ_y$. The operators E_α commute with all of the operators C_μ, since the latter commute with all of the operators of the set. Furthermore, the commutator of an operator E_α with an operator H_i gives always the same operator E_α multiplied by a constant α_i depending upon the particular operators H_i and E_α. The operators E_α are thus step operators which shift the eigenvalue of the operators H_i by an amount α_i. Thus

$$E_\alpha|C'_\mu, H'_i\rangle = K(C'_\mu, \alpha, H'_i)|C'_\mu, H'_i+\alpha_i\rangle \tag{1.21}$$

where the constant $K(C'_\mu, \alpha, H'_i)$ can be determined by algebraic means in a manner similar to the corresponding constants for the angular momentum operators, but which may involve more tedious calculations if there are more operators in the set. Thus eq. (1.21) together with the trivial result

$$H_i|C'_\mu, H'_i\rangle = H'_i|C'_\mu, H'_i\rangle \tag{1.22}$$

define the matrix elements for all of the operators of the set for a complete set of states.

Note that beginning with any particular state $|C'_\mu, H'_i\rangle$, a set of states can be generated by operating successively with the step operators E_α. In this way one can generate sets of states or multiplets. Using the explicit form of the coefficients in eq. (1.21) and limitations on the size of the multiplet, one can arrive at the structure of the multiplet in the same way as one finds that angular momentum multiplets consist of $2J+1$ states having eigenvalues of M changing in steps of one from $-J$ to $+J$.

Diagrams similar to those of Fig. 1.1 can be drawn to represent any Lie algebra and the associated multiplets. However, if there are several operators H_i which are simultaneously diagonal, several quantum numbers are then required to specify the position of a state in the multiplet. In such a case, the diagrams are not one-dimensional as in Fig. 1.1, but r-dimensional where r is the number of simultaneously commuting operators H_i which exist in the set. The Lie algebra is then said to be of rank r. The angular momentum algebra is thus of rank 1.

There are general rules for combining multiplets like there are rules for coupling angular momenta. For any particular set of multiplets describing parts of the system, one can find which multiplets arise in describing the total system, and coefficients analogous to the vector coupling coefficients can be defined for expressing the wave functions of the combined system which belong to a given multiplet. There may also be other divisions of the kinds of multiplets into different groups analogous to the division of angular momentum multiplets into integral and half-integral angular momenta and there may be particularly simple rules like those for angular momenta regarding the combining of multiplets from the same or different sets.

A continuous group of transformations can be defined from these operators by defining infinitesimal transformations in a manner similar to those defined for rotation. Consider, for example, the transformation

$$\psi' = (1 + i\varepsilon X_\varrho)\psi \, . \tag{1.23}$$

A continuous group of transformations can be built up from the infinitesimal transformations generated in this manner by each of the operators of the Lie algebra. Such a continuous group of transformations is called a Lie group. In the conventional treatment, one starts with a continuous group and finds the underlying Lie algebra. We do the reverse beginning with the Lie algebra and we may not even talk about the associated Lie group at all. The results which we use are those summarized in eqs. (1.15)–(1.22). These depend only on the existence of the Lie algebra and do not require the existence of the associated Lie group. We shall also see that physical problems often arise in which the Lie algebra appears naturally in the physical conditions of the problem, while the associated Lie group does not have any simple physical interpretation. The main use for the Lie group in these cases is to provide a convenient label for the Lie algebra and thus indicate where useful studies of this algebra may be found in the literature.

Multiplets of operators or irreducible tensors can be defined for any Lie algebra in a manner analogous to those for angular momentum. One finds sets of operators which can be placed in a one-to-one correspondence with particular multiplets. The commutators of such irreducible tensor operators with the operators E_α and H_i of the Lie algebra are analogous to the corresponding relations (1.21) and (1.22) for the wave function multiplets. The commutator of a particular component of an irreducible tensor operator with a step operator E_α is just the appropriate component of the same irreducible tensor, whereas the commutator with the diagonal operators H_i of any component of an irreducible tensor gives the same component again. Matrix elements of different components of an irreducible tensor operator between two states within the same two multiplets are related by a generalization of the Wigner–Eckart theorem. In the general case, there may be more than a single reduced matrix element required to determine all the matrix elements completely, a situation which does not arise in the angular momentum algebra.

The application of the Lie algebra to a physical problem is directly analogous to the corresponding application of the angular momentum algebra. The application to the solution of the time-independent Schrödinger equation depends upon the form of the commutation relations between the Hamiltonian and the operators of the Lie algebra. Again there are several possibilities:

(1) The Hamiltonian commutes with all the operators. A complete set of eigenstates of the Hamiltonian can be found which are also eigenfunctions of all the operators C_μ and H_i. All the states within a multiplet are degenerate eigenstates of the Hamiltonian. The use of the Lie algebra therefore simplifies the solution of the eigenvalue problem for the Hamiltonian by defining a number of integrals of the motion; i.e. quantum numbers which can be used to specify the eigenstates of the Hamiltonian.

(2) The Hamiltonian does not commute with all of the operators of the algebra but still commutes with the operators C_μ and H_i. One can still define a complete set of eigenfunctions of the Hamiltonian which are also eigenfunctions of these operators but the states within a given multiplet are no longer degenerate.

(3) The Hamiltonian does not commute with all the operators of the Lie algebra but the commutators have a simple form. The Lie algebra is still useful in determining the eigenfunctions and eigenvalue spectrum of the Hamiltonian. Some of the operators of the Lie algebra may be considered as satisfying simple equations of motion and generating elementary excitations of the system.

(4) Any of the properties above apply not to the exact Hamiltonian but to an approximate Hamiltonian which is used as a basis of perturbation theory. The treatment of the unperturbed Hamiltonian is then simplified by the use of the Lie algebra as described above. If the perturbation or 'symmetry-breaking' part of the Hamiltonian is expressed simply in terms of the generalized irreducible tensors, the first-order energy splittings are given by the generalized Wigner–Eckart theorem.

Similar relations are obtainable for the study of transitions from an initial to a final state. If the Hamiltonian commutes with the operators of the Lie algebra and the initial state is an eigenfunction of the operators C_μ and H_i, the final state must also be an eigenfunction of these operators with the same eigenvalues. If the initial state is not an eigenfunction of these operators, it can be expanded in these eigenfunctions. Each non-vanishing term in the expansion then defines a 'channel' through which the reaction can proceed. The most common application is to reactions in which the initial state consists of two particles, an incident particle and a target. Both particles in the initial state may be represented by wave functions which are individually eigenfunctions of the operators C_μ and H_i. The state of the combined system is then an eigenfunction of the operators H_i with an eigenvalue equal to the sum of the two corresponding eigenvalues. However, the state of the combined system is in general not an eigenfunction of the operators C_μ. It is a linear combination of eigenfunctions of C_μ with different eigenvalues in the same way that the product of two angular momentum eigenfunctions is generally not an angular momentum eigenfunction but is some linear combination of angular momentum eigenfunctions. The different eigenvalues of C_μ each define an independent reaction channel which is uncoupled from the others, analogous to angular momentum partial waves.

The relation of the Lie algebra to the physical problem can also be expressed as some symmetry of the Hamiltonian, just as angular momentum algebra is related to the invariance of the Hamiltonian under rotations. A formal symmetry can always be obtained if the operators of the Lie algebra satisfy simple commutation relations with the Hamiltonian. The continuous group of transformations constructed from relations like (1.23) must also transform the Hamiltonian in a simple way. However, these continuous transformations may not have any clear physical meaning, in contrast to the case of rotations. In such cases, the symmetry of physical interest associated with the Lie algebra may be certain discrete transformations. Examples of this type are given in the following chapters.

1.3. PROPERTIES OF BILINEAR PRODUCTS OF SECOND QUANTIZED CREATION AND ANNIHILATION OPERATORS

Let $a_k^\dagger$ and a_k be creation and annihilation operators for a particle in a quantum state k. The pedestrian reader should not be alarmed by the sudden appearance of second quantized field operators since we shall use them only in a very simple way. Consider first operators creating and annihilating bosons. These operators satisfy the commutation rule

$$[a_k, a_k^\dagger] = 1 \tag{1.24}$$

for the annihilation and creation operators of the same quantum state k. Commutators involving two different states all vanish.

We can construct bilinear products of these operators having the form $a_k^\dagger a_m$, $a_k^\dagger a_m^\dagger$ and $a_k a_m$. Note that the commutator of any two such bilinear products is either zero or a linear combination of bilinear products such as, for example, the commutator

$$[(a_k a_m^\dagger), (a_k^\dagger a_n^\dagger)] = a_m^\dagger a_n^\dagger , \tag{1.25}$$

assuming that the states k, m and n are all different. In any such commutator an annihilation operator in one member kills off a creation operator in the other member according to the commutation rule (1.24), thus leaving only the remaining two operators and giving a bilinear product. The commutation rule (1.25) is just the kind of expression that we need to define a Lie algebra. If we have a finite number of states k and construct all possible bilinear combinations of creation and annihilation operators, the commutator of any two bilinear products gives a linear combination of members of the set of bilinear products. A Lie algebra is therefore defined.

It is perhaps surprising to note that a Lie algebra is also defined for bilinear products of fermion creation and annihilation operators. Although it is not the commutation rules of fermion operators which are normally defined but the anticommutation rules, it turns out that these reduce to ordinary commutators where bilinear products are involved. As an example, consider the commutator

$$[a_k , (a_k^\dagger a_m)] \tag{1.26}$$

of the fermion annihilation operators a_k with the bilinear product $a_k^\dagger a_m$. These fermion operators satisfy the anticommutation relations

$$\begin{aligned} a_k a_k^\dagger + a_k^\dagger a_k &= 1 , \\ a_k a_m + a_m a_k &= 0 , \\ a_k^\dagger a_m + a_m a_k^\dagger &= 0 . \end{aligned} \tag{1.27}$$

Using these anticommutation relations we find that the commutator (1.26) can be simplified:

$$[a_k, (a_k^\dagger a_m)] = a_k a_k^\dagger a_m - a_k^\dagger a_m a_k = a_k a_k^\dagger a_m + a_k^\dagger a_k a_m = a_m . \quad (1.28)$$

The commutator of a single fermion operator and a bilinear product of fermion operators is again a single fermion operator or, in general, a linear combination of single fermion operators. Thus the commutator (rather than the anticommutator) of two bilinear products of fermion operators is a linear combination of bilinear products of fermion operators. Such bilinear products also form a Lie algebra if one considers a finite number of states k and all possible bilinear products.

We now see how Lie algebras can arise very naturally in many physical problems. Bilinear products of second quantized creation and annihilation operators can be of interest physically in a wide variety of problems either in field theory or in many-particle systems. Before examining specific cases, let us just make a few further general observations about the kinds of bilinear product which can arise.

We have already noted that there are two types of bilinear products: those referring to boson operators and those referring to fermion operators. The Lie algebras defined by bilinear products of boson operators are simply related to those for fermion operators with a few small differences. The commutators have the same structure, but there may be a difference of sign in some commutators between the boson and the fermion case. The operators $a_m^\dagger a_m^\dagger$ or $a_m a_m$ which either create or annihilate a pair of particles in the same quantum state are perfectly reasonable operators for the boson case. However, for the fermion case, these operators vanish identically because of the Pauli principle (or the anticommutation relations).

The bilinear products can also be divided as follows: There are those like $a_k^\dagger a_m$, the product of a creation and an annihilation operator, which annihilate one particle and create another and therefore do not change the number of particles in the system. There are also those like $a_k^\dagger a_m^\dagger$ or $a_k a_m$ which either create a pair of particles or annihilate a pair of particles and therefore change the number of particles in the system. The commutator of a pair of bilinear products, each of which does not change the number of particles in the system, gives a linear combination of operators which also do not change the number of particles in the system. One can therefore construct Lie algebras containing only those bilinear products which do not change the number of particles. Thus if one considers the most general Lie algebra which can be constructed from a particular finite set of creation and annihilation operators, one finds that this includes operators of both types: those which change the number of particles, and those which do not. Another Lie algebra is formed by a subset of these operators consisting of all operators of the set which do not change the number of particles.

CHAPTER 2

ISOSPIN. A SIMPLE EXAMPLE

2.1. THE LIE ALGEBRA

The simplest case of a Lie algebra generated from bilinear products of creation and annihilation operators is the case where there are only two quantum states. This is just the case of 'old-fashioned' isopin as it was originally conceived for systems of neutrons and protons before the discovery of mesons and strange particles. Let $a_p^\dagger$ and $a_n^\dagger$ be operators for the creation of a proton and a neutron, respectively, and let a_p and a_n be the corresponding annihilation operators. For the present we do not consider the space and spin states of these particles and assume that there is only one quantum state for the proton and one quantum state for the neutron. We shall put in the space and spin later.

Let us now construct the Lie algebra of all possible bilinear products of these operators which do not change the number of particles. These are products of one creation operator and one annihilation operator. Since there are two possible operators of each type, there are in all four possible bilinear products which do not change the number of particles:

$$a_p^\dagger a_n, \qquad a_n^\dagger a_p, \qquad a_p^\dagger a_p \quad \text{and} \quad a_n^\dagger a_n .$$

The first of these operators annihilates a neutron and creates a proton in the same quantum state; i.e. it changes a neutron into a proton. The second operator does the reverse, changing a proton into a neutron. These two operators are thus the ordinary isospin operators τ_+ and τ_-. The other two operators annihilate either a proton or a neutron and create the same particle back again. These are just number operators which count the number of protons and neutrons. The sum of the last two operators is just the total number operator which counts the number of nucleons. Since all of the other operators do not change the number of nucleons, this total number operator commutes with all the others. It is therefore convenient to divide the set of four operators into a set of three plus the total number, or baryon number, operator which commutes with all of the others.

$$B = a_p^\dagger a_p + a_n^\dagger a_n \,, \tag{2.1a}$$

$$\tau_+ = a_p^\dagger a_n \,, \tag{2.1b}$$

$$\tau_- = a_n^\dagger a_p \,, \tag{2.1c}$$

$$\tau_0 = \tfrac{1}{2}(a_p^\dagger a_p - a_n^\dagger a_n) = Q - \tfrac{1}{2}B \,. \tag{2.1d}$$

The operator τ_0 defined as half the difference between the number of protons and the number of neutrons is just equal to the total charge Q minus half the baryon number, since the protons carry one unit of charge and the neutrons carry no charge. The operators τ_+, τ_- and τ_0 satisfy commutation rules exactly like angular momenta

$$[\tau_0, \tau_+] = \tau_+ \,, \tag{2.2a}$$

$$[\tau_0, \tau_-] = -\tau_- \,, \tag{2.2b}$$

$$[\tau_+, \tau_-] = 2\tau_0 \,. \tag{2.2c}$$

This has led to the designation isospin for these operators and to the description of rotations in a fictitious isospin space. Since the commutation rules for the isospin operators are exactly the same as those for angular momenta, we can immediately take over all of the results of angular momentum which follow from the commutation rules. However, let us first put in the space and spin degrees of freedom for the proton and neutron. Let $a_{pk}^\dagger$ represent the creation operator for a proton in a quantum state k where the letter k indicates both the space and the spin state, and similarly for neutrons and for the annihilation operators. The isospin operators defined by eq. (2.1) can now easily be generalized simply by adding the index k everywhere and summing over the index k

$$B = \sum_k a^\dagger_{pk} a_{pk} + a^\dagger_{nk} a_{nk} , \tag{2.3a}$$

$$\tau_+ = \sum_k a^\dagger_{pk} a_{nk} , \tag{2.3b}$$

$$\tau_- = \sum_k a^\dagger_{nk} a_{pk} , \tag{2.3c}$$

$$\tau_0 = \tfrac{1}{2}\sum_k (a^\dagger_{pk} a_{pk} - a^\dagger_{nk} a_{nk}) = Q - \tfrac{1}{2}B . \tag{2.3d}$$

The commutation rules (2.2) are also valid for the new definitions (2.3) of the isospin operators. Since bilinear products corresponding to two different quantum states k and k' commute, the only terms in any commutator which give a non-vanishing contribution are those which refer to a single quantum state k. Thus, each quantum state k acts independently in the commutator and the operators obtained are always a sum over all values of k. Since the summation over the space-spin indices does not affect the Lie algebra, the simpler notation of eq. (2.1) is used from this point, with the understanding that this is a shorthand for writing down the more cumbersome expressions (2.3) involving sums over the space-spin indices.

By analogy with angular momentum we see that there exists an operator T^2 analogous to J^2, which commutes with all the operators (2.3). States of a neutron–proton system can be classified into multiplets, each characterized by an eigenvalue of the operator T^2. Each state can be chosen to be a simultaneous eigenfunction of τ_0 and T^2. The eigenvalues of T^2 have the form $T(T+1)$ where T is either an integer or a half-integer. Each multiplet consists of $2T+1$ states with eigenvalues T_0 of the operator τ_0 varying in the steps of unity from $-T$ to $+T$.

The relation (2.3d) expressing the operator τ_0 in terms of the charge and baryon number indicates independently that this operator can have only integral or half-integral eigenvalues. Such considerations can be useful in other cases where the result is not already known by the direct analogy with angular momentum. We also obtain certain rules for combining multiplets directly by noting that the total charge and the total baryon number are additive quantum numbers and therefore τ_0 also is additive. Thus if a system consists of several parts, each of which is in a state which is an eigenfunction of τ_0, the whole system is described by a state which is also an eigenfunction of τ_0 and the eigenvalue is simply the sum of the eigenvalues of the separate parts. Thus if two multiplets having integral values of T_0 are combined, the multiplet describing the overall system must also have an integral value of T_0, and similarly for two multiplets having half-integral values of T_0. On the other hand, if a multiplet having an integral value of T_0 is combined with one having a half-integral value the multiplet describing the combined system then must have half-integral values of T_0.

Let us now consider which Lie group is associated with these isospin operators. By analogy with the angular momentum operators, we allow these operators to generate infinitesimal transformations such as

$$\psi' = \{1 + i\varepsilon(\tau_+ + \tau_-)\}\psi \,. \tag{2.4}$$

We use the linear combination $\tau_+ + \tau_-$ because these operators individually are not hermitean. Note that such a transformation changes a proton or a neutron into something which is a linear combination of the proton and neutron state. These transformations are thus transformations on complex vectors in a two-dimensional proton–neutron Hilbert space. The transformations are unitary; thus the Lie group associated with isospin is some group of unitary transformations in a two-dimensional space.

The whole group of unitary transformations in a two-dimensional space is generated by the set of four operators (2.3) including the baryon number B. The unitary transformations generated by the operator B are of a very trivial nature, namely, multiplication of any state by a phase factor. Since the three isospin operators form a Lie algebra by themselves, the associated continuous group is a subgroup of the full unitary group in two dimensions. This group is usually called the special unitary group or unimodular unitary group and denoted by the letters SU_2. This is the group of unitary transformations which are represented by the matrices having a determinant of $+1$. Such transformations clearly form a group by themselves since the product of any two matrices having a determinant of $+1$ is also a matrix having a determinant of $+1$.

Thus isospin transformations are two-dimensional unitary transformations rather than three-dimensional rotations. There is no three-dimensional space which has any direct physical interpretation. The analogy with angular momentum is purely formal, and arises because the Lie algebra of operators generating unitary transformations in a two-dimensional space happens to be the same as the algebra of the operators generating rotations in a three-dimensional space.

Operators satisfying commutation rules like angular momenta mysteriously arise in a number of physical problems. These are often called 'quasispins' but have no direct physical interpretation in terms of any rotation in a real three-dimensional space. The reason why such quasispins often occur is that there is only one Lie algebra of rank 1, where the states within a multiplet are specified completely by one quantum number and where the diagram of the multiplet as shown in Fig. 1.1 is a one-dimensional plot. This algebra is just the angular momentum algebra. Thus no matter what kind of transformation is being considered, rotations, unitary transformations, or more complicated ones such as symplectic transformations (cf. § 5.4), these all give a Lie algebra which is the same as the angular momentum algebra when the number of dimensions in the space where the transformations are taking place is sufficiently small so that the Lie algebra must be of rank 1. There is no real physical three-dimensional space associated with these quasispins. The descriptions in terms of rotations in quasispin or isospin space are purely formal, have no direct physical meaning, and are useful only because we are familiar with the algebraic properties of angular momentum operators.*

2.2. THE USE OF ISOSPIN IN PHYSICAL PROBLEMS

The algebra of isospin operators is useful in physical problems involving nucleons because of the charge independence of nuclear forces. This implies that the Hamiltonian describing nuclear forces commutes with the three isospin operators, (2.3), and that states of nucleons can be classified into multiplets characterized by the value of the total isospin quantum number T. There are, of course, electromagnetic forces which are not charge independent. However, these are weak in comparison with the nuclear forces and can be considered as a perturbation. The different states of an isospin multiplet are therefore not degenerate; there is a small splitting due to electromagnetic effects.

The isospin Lie algebra takes on added significance when mesons and strange particles are introduced. These are easily incorporated into the isospin scheme. All the results regarding isopin multiplets and matrix elements of operators follow from the commutation rules (2.2) and do not depend upon the specific definition of the isospin operators (2.3) in terms of neutrons and protons. The new particles are therefore included in the isospin scheme simply by requiring that they fit into isospin multiplets characterized by a particular value of T and that there are $2T+1$ states in a multiplet with eigenvalues T_0 of the operator τ_0 which vary in steps of unity from $-T$ to $+T$. The isospin operators are now no longer defined

* One may note here that the continuous group of isospin transformations is very peculiar since they transform physical nucleon states into states which contain linear combinations of neutrons and protons. Such linear combinations are never observed physically because of charge conservation and it has been suggested that such states do not exist in the Hilbert space describing physical states because of superselection rules. It is therefore perhaps satisfying that all of the useful isospin results can be obtained directly from the Lie algebra which involves only physical operators acting upon physical states and that the unphysical Lie group of continuous transformations is not required in order to obtain any of these results.

by eqs. (2.3) which apply only to nucleons, but can be completely defined by the relations which give the matrix elements of the operators within a particular multiplet.

$$\tau_\pm|T, T_0\rangle = \sqrt{T(T+1)-T_0(T_0\pm 1)}\,|T, T_0\pm 1\rangle \qquad (2.5a)$$

$$\tau_0|T, T_0\rangle = T_0|T, T_0\rangle. \qquad (2.5b)$$

A typical multiplet is shown in Fig. 2.1.

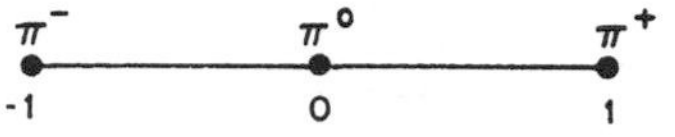

Fig. 2.1

The relations (2.3d) between the operator τ_0 and the charge and baryon number no longer hold since these refer only to nucleon systems. However, the rules for combining multiplets having integral or half-integral eigenvalues of T_0 remain valid. They are a general property of the Lie algebra and do not depend specifically on the assumption that all states are made up of neutrons and protons.

Let us now examine some of the physical implications and consequences of extending the isospin formalism to particles other than nucleons. We first note that the experimentally observed charge independence of nuclear forces requires that the interactions in any system of nucleons are not changed under the isospin transformations which transform neutrons and protons into one another. This implies that any particles which interact strongly with nucleons must also transform in a simple way into other particles having different charges. Otherwise the contribution of these particles to the nuclear forces would change when protons are changed into neutrons and the overall charge independence of nuclear forces would be violated. The incorporation of all strongly interacting particles into the isospin formalism therefore follows from the experimentally observed charge independence of nuclear forces.

The extension of isospin to other strongly interacting particles leads to the following types of experimental predictions:

A. Predictions from the multiplet structure

1. *Classification.* The most obvious prediction is that all new strongly interacting particles which are found and all resonances

observed between strongly interacting particles must belong to isospin multiplets. Once one member of a given multiplet is found, all the other members of the multiplet must also exist.

2. *Couplings.* If one considers resonances between two particles, the multiplet structures which can arise are determined by the isospin coupling rules. For example, all resonances of pions ($T=1$) and nucleons ($T=\frac{1}{2}$) must belong to multiplets having either $T=\frac{1}{2}$ or $T=\frac{3}{2}$.

B. Predictions of relations between matrix elements

1. *Decays.* The decay rates or widths of resonances belonging to the same isospin multiplet are related by isospin coupling rules. Consider, for example, a nucleon–pion resonance having $T=\frac{3}{2}$. There are four charge states for this isospin multiplet and six possible decay modes, since all charge states of the nucleon–pion system are possible final states for the decay. The transition matrix elements for all six decay modes are proportional to one another; i.e. they are all proportional to the same reduced matrix element with a proportionality factor which is a Clebsch–Gordan coefficient for the coupling of $T=\frac{1}{2}$ to $T=1$ to give a total $T=\frac{3}{2}$.

2. *Reactions.* Cross sections for different reactions involving members of the same multiplets are related in a manner involving isospin coupling rules (usually involving some vector addition or Clebsch–Gordan coefficients). For example, if one is considering pion–nucleon scattering including both elastic and charge exchange processes, one notes that there are three pion states and two nucleon states and thus six possible elastic scattering reactions. There are also two independent charge exchange reactions ($\pi^+ \text{n} \to \pi^0 \text{p}$ and $\pi^- \text{p} \to \pi^0 \text{n}$) giving a total of eight. On the other hand, any pion–nucleon state can be expressed as a linear combination of members of a multiplet having $T=\frac{1}{2}$ and a multiplet having $T=\frac{3}{2}$. Since the interaction responsible for the scattering conserves isospin, there are only two independent channels, and all the eight processes considered should have their cross sections expressible in terms of two complex amplitudes, the $T=\frac{1}{2}$ amplitude and the $T=\frac{3}{2}$ amplitude. Thus eight experimental cross sections are determined by

three real parameters: the magnitudes of the two scattering amplitudes and the relative phase. This result can be expressed as predictions of relations between the various elastic and charge exchange scattering cross sections of pions and nucleons. Similar relations exist for all reactions involving strongly interacting particles.

3. *Selection rules.* One can also find selection rules which result from isospin. For example, if one considers only nucleons and pions, states having an odd baryon number must belong in a multiplet with a half-integral isospin, whereas states with an even baryon number must belong in a multiplet having an integral isospin. From this we obtain a general selection rule: a state that does not satisfy these conditions cannot decay into any combination of nucleons and pions by strong interactions which are invariant under the isospin transformations (i.e. whose Hamiltonian commutes with the operators (2.3)). The Σ-hyperon with odd baryon number and integral isospin and the K-meson with baryon number zero and half-integral isospin are examples of this selection rule and they can only decay by weak interactions in which isospin is not conserved.

C. *Symmetry-breaking effects*

The above predictions all follow from the assumption that the Hamiltonian describing strong interactions is invariant under isospin transformations; i.e. it commutes with the isospin operators. The isospin formalism is useful in making predictions also for the case where the Hamiltonian is not invariant under isospin transformations, if its transformation properties can be expressed in a simple way. This is the case for the electromagnetic interaction which is not charge independent and which does not commute with the isospin operators τ_+ and τ_-. The electromagnetic interaction behaves under isospin transformations like a linear combination of an isoscalar and an isovector. This can be seen by noting that within any isospin multiplet the electric charge of a particle is the sum of a constant and the eigenvalue T_0. The constant commutes with all the isospin operators. T_0 is the eigenvalue of an operator

τ_0 which is a member of an isospin triplet; i.e. it behaves like a member of an isospin multiplet with $T=1$.

This transformation property is particularly useful in cases where the electromagnetic interaction can be treated as a perturbation. Let us consider the operation of the electromagnetic interaction on a particular state $|T, T_0\rangle$

$$E|T, T_0\rangle = (I_s + I_v)|T, T_0\rangle \tag{2.6}$$

where I_s and I_v are the isoscalar and isovector parts, respectively, of the electromagnetic interaction E. We first note the following selection rule: The isoscalar part of the electromagnetic interaction commutes with all of the isospin operators and therefore cannot change the eigenvalues of T and T_0. The isovector part behaves like an element of a $T=1$ multiplet which is coupled by ordinary angular momentum coupling rules. Thus I_v has nonvanishing matrix elements only between the state $|T, T_0\rangle$ and states of total isospin $T+1$, T and $T-1$ and has no matrix element connecting two $T=0$ states. We also note that the matrix elements of I_v between different pairs of states in the same isospin multiplets are related by the Wigner–Eckart theorem.

If we are considering a reaction which goes via first-order perturbation theory in the electromagnetic interaction, the transition probability depends upon the matrix elements of the electromagnetic interaction between the initial and final states. We thus obtain selection rules and relations between reactions involving members of the same isospin multiplets.

If the reaction considered does not go by first-order perturbation theory, more complicated relations are obtained. If we are considering an nth-order process, the electromagnetic interaction operator acts n times. If we wish to consider the corresponding isospin coupling we must consider all possible couplings of n isovectors as well as the replacement of some of these isovectors by isoscalars. The situation in the general case is therefore so complicated that useful predictions are rarely obtained. However, since many radiative processes are of first order useful predictions of the kind described above can often be obtained.

2.3. THE RELATION BETWEEN ISOSPIN INVARIANCE AND CHARGE INDEPENDENCE

Although the invariance of strong interactions under isospin transformations is synonymous with the charge independence of forces between nucleons, the forces between other kinds of particles are *not necessarily* charge independent. It should be emphasized that isospin invariance does not, for example, require that forces between pions be charge independent. To see this let us first examine how isospin invariance implies the charge independence of forces between nucleons and we shall see that the same arguments are not valid for pions.

Consider the interaction between two nucleons in a state which is antisymmetric in space and spin. Such a space-spin state has three possible charges: it can be a two-proton state, a two-neutron state, or a proton–neutron state. The isospin formalism says that these three states form an isospin multiplet with $T=1$. If the interactions are invariant under isospin transformations; i.e., they commute with the isospin operators, then the interaction must be the same in every state of the multiplet. The proton–proton, proton–neutron and neutron–neutron interactions are thus all the same in states which are antisymmetric in space and spin. For states which are symmetric in space and spin there is no argument since such states can only be neutron–proton states with no possibility of other charges.

Let us now consider the interaction of two pions in a state which is symmetric in space (no spin). Such a state has six possible charge states: (π^+, π^+), (π^-, π^-), (π^0, π^0), (π^+, π^0), (π^0, π^-) and (π^-, π^+). The isospin formalism says that two pions in a symmetric spatial state can have either $T=0$ or $T=2$. In other words, the six space-symmetric charge states of two pions form two isospin multiplets: a quintet and a singlet. If strong interactions are invariant under isospin transformations, then the interaction between two pions must be the same for any state within a given multiplet. However, the isospin invariance makes no prediction of the relation between the interaction of two pions in the $T=2$ state and in the $T=0$ state. Thus the interaction of two pions is not the same in all six possible charge states. The two neutral states, (π^+,π^-) and (π^0,π^0) are both linear combinations of the two isospin states, $T=2$ and $T=0$, and the interaction is therefore determined by the two parameters specifying the interactions in these two states. Thus, *isospin invariance does not require that the forces between two pions be charge independent.*

We see that isospin invariance requires charge independence only for the forces between pairs of particles which form an isospin doublet with $T=\frac{1}{2}$ like the nucleons. For all higher multiplets the forces are in general not charge independent. To require these forces to be charge independent requires an additional greater symmetry beyond that of isospin.*

Let us now consider the interaction between Σ-hyperons and pions and the restrictions imposed by isospin invariance. There are three charge states of each, and thus in all nine possible charge states for the Σ–π system. Since these are two different particles, there are no restrictions imposed by the Pauli principle. Both the Σ and the π have isospin 1, thus the nine states of the Σ–π system are distributed among three isospin multiplets: a quintet having $T=2$, a triplet having $T=1$, and a singlet having $T=0$. If the Hamiltonian is invariant under isospin transformations, the interaction must be the same for all Σ–π states within the same multiplet. Thus, the interaction between Σ's and π's in the nine possible charge states are expressed as functions of three parameters; the quintet interaction, the triplet interaction, and the singlet interaction. However, isospin invariance does not require any relation between these three interactions. Thus, the Σ–π interaction is not charge independent in the sense that it is independent of the charges of the Σ and the π. However, there are relations between the interactions in that there are only three independent interactions instead of nine. The basis of these relations is easily understood in terms of the requirement that the interaction between nucleons be completely charge independent and that the interactions between other particles be

* One can see that complete charge independence for pion–pion forces would require that the singlet and quintet interactions be equal.

restricted in the manner required to maintain the nucleon charge independence.

One might imagine a situation where the Σ's and the π's were the first particles to be discovered and nucleons for some reason did not exist or were unstable. This rather artificial situation is considered here because the analogous situation does exist in the octet or eight-fold model of elementary particles with unitary symmetry. One would then find experimentally that there were relations between the interactions in the different Σ–π charge states which were described simply in terms of the algebra of the group SU_2. One might say that these interactions behaved as if there existed a doublet of basic particles for which the interactions were really charge independent. On the other hand, one could also give a simple description of the Σ–π interaction in terms of the isospin operators without requiring that the nucleon or some other isospin doublet exist.

2.4. THE USE OF THE GROUP THEORETICAL METHOD

The simple example of isospin illustrates the use and the power of the group theoretical method. In this case it was not even necessary to investigate the algebra of the operators or the structure of the multiplets. All that was necessary was to show that the operators satisfied the same commutation rules as angular momentum operators. From this point it was possible to use all of the results already known from angular momenta, even though the physical situation described by isospin was very different from rotations in ordinary three-dimensional space. One finds repeatedly in physics that abstract algebraic relations obtained in one kind of physical problem can be useful in another problem where the same algebra arises.

We note again that the physical basis of isospin is the charge independence of nuclear forces and the coupling of all strongly interacting particles in accordance with this charge independence. The isospin formalism does not add any new physics to this basis. It merely offers a simple and compact method for calculating the consequences of this basic physical principle for experimental measurements on strongly interacting particles.

In this example we have seen how Lie algebra can be generated from bilinear products of creation and annihilation operators and how this algebra is useful in two cases: (1) The strong interactions, where the Hamiltonian commutes with all of the isospin operators; (2) The electromagnetic interaction which has simple commutation relations with the isospin operators and which is sufficiently small to be treated as a perturbation. We did not need to investigate the structure of the algebra and the multiplets in detail because these were immediately evident by the connection with angular momentum. In the following chapter we consider a more complicated example where the algebra and multiplet structure must be investigated, but the general treatment is a simple extension of isospin.

CHAPTER 3

THE GROUP SU_3 AND ITS APPLICATION TO ELEMENTARY PARTICLES

3.1. THE LIE ALGEBRA

We have seen that the isospin Lie algebra is generated from bilinear products of creation and annihilation operators in the case where there are only two quantum states. Consider now the case where there are three quantum states. A convenient example of this case is the Sakata model of elementary particles in which the transformations of isospin are extended to include the lambda hyperon as well as the proton and the neutron. Let $a_\Lambda^\dagger$ and a_Λ be operators for the creation and annihilation of a lambda particle. We now construct the Lie algebra of all possible bilinear products of the nucleon and lambda operators which do not change the number of particles. With three creation operators and three annihilation operators, there are nine possible bilinear products. These are conveniently written as follows:

$$\begin{aligned}
B &= a_\mathrm{p}^\dagger a_\mathrm{p} + a_\mathrm{n}^\dagger a_\mathrm{n} + a_\Lambda^\dagger a_\Lambda\,, \\
\tau_+ &= a_\mathrm{p}^\dagger a_\mathrm{n}\,, \qquad \tau_- = a_\mathrm{n}^\dagger a_\mathrm{p}\,, \\
\tau_0 &= \tfrac{1}{2}(a_\mathrm{p}^\dagger a_\mathrm{p} - a_\mathrm{n}^\dagger a_\mathrm{n})\,, \\
B_+ &= a_\mathrm{p}^\dagger a_\Lambda\,, \qquad B_- = a_\mathrm{n}^\dagger a_\Lambda\,, \\
C_+ &= a_\Lambda^\dagger a_\mathrm{n}\,, \qquad C_- = a_\Lambda^\dagger a_\mathrm{p}\,, \\
N &= \tfrac{1}{3}(a_\mathrm{p}^\dagger a_\mathrm{p} + a_\mathrm{n}^\dagger a_\mathrm{n} - 2a_\Lambda^\dagger a_\Lambda) = \tfrac{1}{3}B + S\,.
\end{aligned} \tag{3.1}$$

As in the case of isospin, we are not writing sums over space and spin variables, but are using the simpler notation of eq. (2.1).

As in the case of isospin, some of the bilinear products are operators which change one kind of particle into another, while others are number operators which simply count the number of particles of a particular kind. Again the sum of all the number operators is just the baryon number and commutes with all of the other operators which do not change the baryon number. We therefore divide the set of nine operators into a set of eight plus the baryon number which commutes with all of the rest. In the set of eight operators, there are still two number operators and it is convenient to choose the linear combinations given in eq. (3.1); namely, the isospin operator τ_0 and the operator N which is just one-third the difference between the number of nucleons and twice the number of lambdas. Since the nucleons have strangeness 0 and the lambda strangeness -1, the operator N is just equal to the sum of one-third of the baryon number and the strangeness as indicated in eq. (3.1). The remaining six operators in the set are the two isospin operators τ_+ and τ_- and the four operators B_+, B_-, C_+ and C_- which change lambdas into nucleons and vice versa.

Let us now consider which Lie group is associated with these operators. By an extension of isospin we see that these operators generate infinitesimal transformations in a three-dimensional proton–neutron–lambda Hilbert space. These transformations are again unitary; thus the Lie group associated with these operators is the group of unitary transformations in a three-dimensional space. Again the full unitary group in three dimensions is generated by the set of nine operators including the baryon number. The set of eight operators excluding the baryon number generates the unimodular unitary group in three dimensions which is designated by the notation SU_3. These are again unitary transformations which are represented by the matrices having a determinant of $+1$.

Inspection of the set of operators (3.1) shows that the eight operators form a Lie algebra of rank 2. The two operators τ_0 and N commute with one another, and it is impossible to find a third operator which commutes with both of these. A few simple observations show that the remaining six operators are already in the desired form of step operators, E_α, defined in eqs. (1.18), (1.19) and (1.20), shifting the eigenvalues of τ_0 and N. The eigenvalue of τ_0 is unaffected by the creation or annihilation of a lambda, which has isospin zero. A change in the eigenvalue of N is the same as a change in strangeness, since none of the operators change the baryon number. Strangeness is unaffected by the creation or anni-

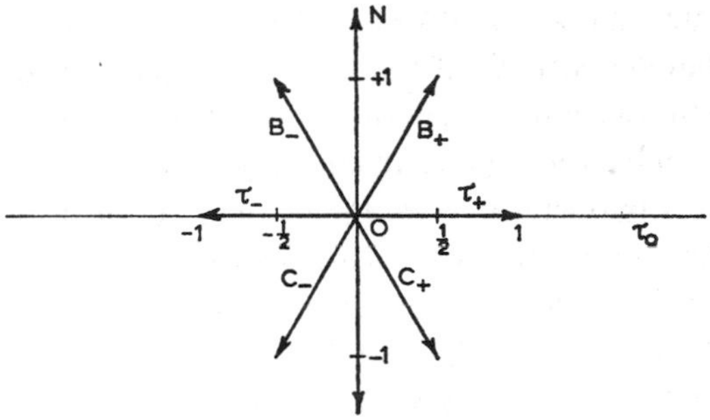

Fig. 3.1

hilation of a nucleon, which has strangeness zero. The operators with the subscript $+$, B_+ and C_+, create a proton and annihilate a neutron, respectively, thereby increasing the eigenvalue of τ_0 of any state by $+\frac{1}{2}$. Similarly, the operators with the subscript $-$ change the eigenvalue of τ_0 by $-\frac{1}{2}$. The B-operators annihilate a lambda and therefore increase the strangeness and the eigenvalue of N by $+1$; the C-operators create a lambda and therefore change the strangeness and the eigenvalue of N by -1. The following commutation relations can thus be written down without any calculation.

$$\begin{aligned} [\tau_0, \tau_\pm] &= \pm\tau_\pm , & [N, \tau_\pm] &= 0 , \\ [\tau_0, B_\pm] &= \pm\tfrac{1}{2}B_\pm , & [N, B_\pm] &= B_\pm , \\ [\tau_0, C_\pm] &= \pm\tfrac{1}{2}C_\pm , & [N, C_\pm] &= -C_\pm . \end{aligned} \tag{3.2a}$$

These commutation rules can be represented on a diagram analogous to Fig. 1.1a for angular momentum. However, since this algebra is of rank 2 the diagram is a two-dimensional plot of the eigenvalues of τ_0 and N as shown in Fig. 3.1.

The remaining commutators are easily obtained by simple algebra.

$$
\begin{aligned}
&[\tau_\pm, B_\pm] = [\tau_\pm, C_\pm] = 0 = [C_+, C_-] = [B_+, B_-]\,, \\
&[\tau_\pm, B_\mp] = B_\pm\,, \qquad [B_\pm, C_\pm] = \tau_\pm\,, \\
&[\tau_\pm, C_\mp] = -C_\pm\,, \qquad [\tau_+, \tau_-] = 2\tau_0\,, \\
&[B_+, C_-] = \tfrac{1}{2}(3N+2\tau_0)\,, \\
&[B_-, C_+] = \tfrac{1}{2}(3N-2\tau_0)\,.
\end{aligned}
\tag{3.2b}
$$

Following the analogy with isospin we might attempt to find operators C_μ which are functions of these operators and commute with all of them. The eigenvalues of these operators would then be used to label the multiplets as the eigenvalues of the operator T^2 label the isospin multiplets. However, the operators C_μ for the SU_3 group are rather complicated and we defer considering them to a later point. We shall see that a considerable amount can be learned about the structure of the multiplets without knowing the explicit form of the operators C_μ.

3.2. THE STRUCTURE OF THE MULTIPLETS

The SU_3 multiplets are generated by successive operation on any state within the multiplet with the eight operators of the Lie algebra represented in Fig. 3.1. The states of each multiplet are represented as points on a two-dimensional plot of the eigenvalues of τ_0 and N analogous to the one-dimensional plot of isospin multiplets in Fig. 2.1. The points representing the states of a given multiplet should appear in such a plot as a two-dimensional lattice in which the lattice vectors are just the vectors of Fig. 3.1 representing the operation of the operators $B_\pm$, $C_\pm$ and $\tau_\pm$. The lattice therefore has the hexagonal character of Fig. 3.1 in which a change of N by ± 1 is always accompanied by a change in T_0 of $\pm\frac{1}{2}$. Since the SU_3 algebra is larger than the isospin algebra and includes it as a subset, we expect the multiplets for SU_3 to be larger than isospin multiplets and to contain several isospin multiplets at different values of N. Since the B- and C-operators change T_0 by $\pm\frac{1}{2}$, both integral and half-integral isospins occur in the same multiplet in contrast to the case of isospin multiplets. From the hexagonal character of the lattice we see that integral and half-integral isospin multiplets appear alternately with increasing values of the quantum number N.

Let us now consider some simple examples of SU_3 multiplets. The neutron–proton–lambda triplet itself forms a multiplet since the eight operators simply transform these particles into one another. This triplet is represented in the diagram, Fig. 3.2a. Similarly, the corresponding antiparticle triplet forms a multiplet and is illustrated in Fig. 3.2b. For convenience, we use the term *sakaton* to denote neutron, proton or lambda analogous to the term nucleon for neutron and proton. Let us now examine the states of the system formed by one sakaton and one antisakaton. The nine states formed by combining these two triplets can be analyzed to determine their behavior under the transformations generated by the operators (3.1). However, since these sakaton–antisakaton states are generated from the vacuum by the operation of a product of a sakaton creation and a sakaton annihilation operator, e.g. $a_p^\dagger a_\Lambda|0\rangle$ we see that these nine states look very much like the nine operators (3.1). If we write down these states explicitly we find that they split into two multiplets, a singlet and an octet analogous to the operators (3.1). These multiplets are shown in Figs. 3.3a and 3.3b. The singlet has $N=0$ and $T_0=0$ and thus has the same SU_3 quantum numbers as the vacuum. The octet looks very much like the diagram, Fig. 3.1, of the generators of the group.

Note that there are two points in the octet of Fig. 3.3b at $N=0$, $T_0=0$. This degeneracy is a characteristic of the SU_3 multiplet which is not found in isospin and angular momentum multiplets. In the latter the eigenvalue of J_z or T_0 is sufficient to specify a state completely within a multiplet. In the SU_3 multiplets the eigenvalues of the two operators N and τ_0 are not always sufficient for complete specification of a state within a multiplet; there may be several states having the same values of these quantum numbers. An additional quantum number is therefore necessary to distinguish between these states. The choice of this additional quantum number is not determined by the algebra of the group SU_3. In general, such additional quantum numbers are chosen for convenience in the particular physical problem under consideration, and the choice is not unique.

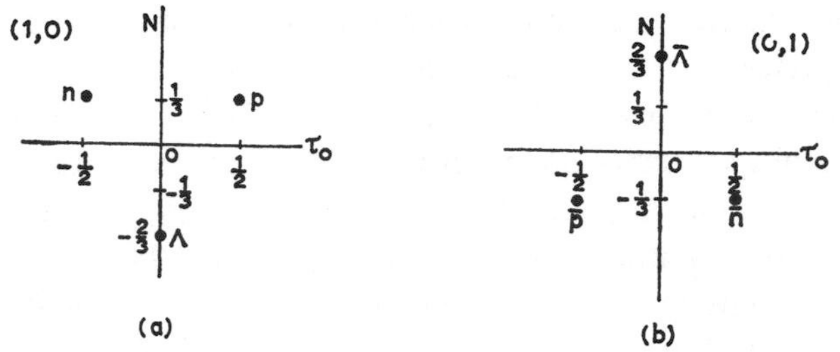

Fig. 3.2

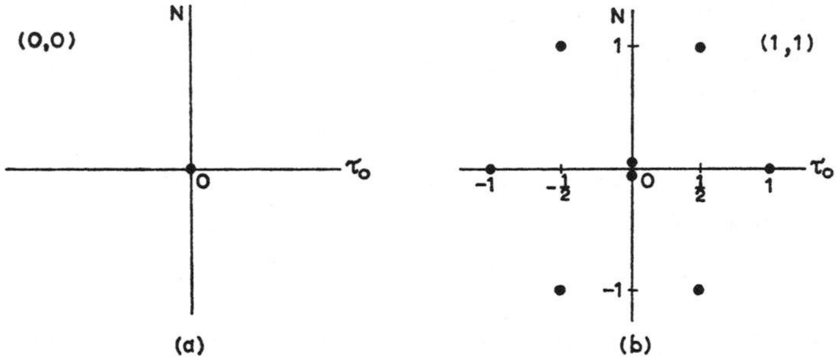

Fig. 3.3

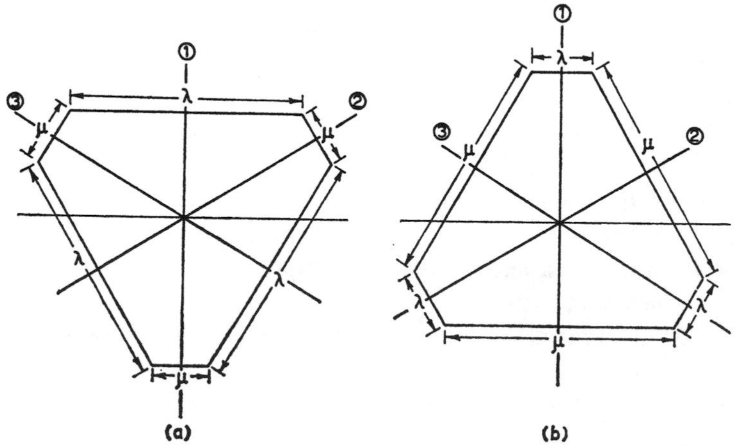

Fig. 3.4

One way to find such additional quantum numbers is to examine other groups which are subgroups of the one being considered. The Casimir operators C_μ of the subgroups may define a convenient additional quantum number. This turns out to be the case in our present analysis of SU_3. The obvious subgroup is the SU_2 group which for physical reasons we should like to keep as a good quantum number. It turns out that the total isospin operator T^2 gives us an additional quantum number which is all that is required for complete specification of the states in a multiplet. In the particular multiplet under consideration, Fig. 3.3b, we see that the four states having $N=0$ split into an isospin triplet and an isospin singlet. Thus the two states having $N=0$ and $T_0=0$ are distinguished by the eigenvalue of T, which is $T=1$ for the triplet state and $T=0$ for the singlet state.

From examination of Figs. 3.2 and 3.3 we see that these multiplets are indeed hexagonal, two-dimensional lattices. To give a complete specification of a multiplet we need to know its shape and size and the number of states appearing at each lattice point. A few simple considerations discussed below show that the shape of the multiplet must always be a hexagon or truncated triangle specified by two parameters as shown in Fig. 3.4. For reasons which will become apparent below these two parameters are called λ and μ and represent the size of the isospin multiplets occurring at the maximum and minimum values of N in the particular SU_3 multiplet.

$$\lambda = 2T \quad \text{at} \quad N = N_{\text{max}} ,$$
$$\mu = 2T \quad \text{at} \quad N = N_{\text{min}} .$$

The multiplets are then labeled (λ, μ). In this notation the two triplets of Fig. 3.1 are denoted by the values (1,0) and (0,1) respectively; the singlet of Fig. 3.3a is denoted by (0,0) and the octet of Fig. 3.3b is denoted by (1,1).

The requirement that the multiplets must have the form of Fig. 3.4 can be deduced from certain simple symmetries and arguments similar to those used in obtaining properties of crystal lattices. We first note that all multiplet diagrams must be symmetric about a vertical axis through the center since a reflection across this axis simply interchanges the neutron and proton in the sakaton multiplet and in general takes any state into the corresponding one of the same isospin multiplet having the equal and opposite eigenvalue of τ_0. (This is just the 'charge symmetry' transformation which is used in defining G-parity.) Since the neutron, proton and lambda are all considered on an equal basis in the Sakata model, the transformations which interchange the proton and lambda leaving the neutron unchanged, or which interchange the neutron and lambda and leave the proton unchanged, should be similar in nature to the proton–neutron transformation discussed above. Thus we see that the diagrams must also be symmetric with respect to reflections about the axes denoted by the numbers 2 and 3 in Fig. 3.4 which are at angles of 120° with respect to the vertical. Cyclic permutations of the n–p–Λ triplet corresponding to rotations of 120° of the multiplet diagram are obtained from successive reflections across two of the axes mentioned above. The multiplet must therefore also have a shape which is invariant under rotations of $\pm 120°$. These symmetry properties are almost sufficient to fix the shape of the multiplet as that given in Fig. 3.4. (The only other possibilities not yet excluded involve re-entrant corners in the polygon.)

We also note that the operation of charge conjugation changes the signs of the quantum numbers N and T_0. The charge conjugates of the particles in a given multiplet such as that shown in Fig. 3.4a then form a multiplet having a shape obtained by inversion through the origin as shown in Fig. 3.4b. The charge conjugate multiplet thus has the values of λ and μ interchanged (e.g., see sakaton and antisakaton multiplets of Fig. 3.2).

Detailed analyses of the properties of the multiplets should include: (1) a rigorous demonstration that the multiplets do indeed have the shape shown in Fig. 3.4; (2) a prescription for the number of states occurring at each lattice point in the multiplet for the cases where several isospin multiplets occur with the same N-value, and (3) explicit expressions for the matrix elements of the operators (3.1) between different states of the same multiplet. This analysis can be carried forward in a variety of ways.

One method would be to follow a procedure analogous to that

used in angular momentum; namely, to obtain relations between matrix elements by use of the commutation rules and by noting that certain operators must have vanishing matrix elements when operating on states at the edge of the multiplet. This method is perfectly straightforward, but the algebra is more complicated and tedious than for the case of angular momentum.

Another line of approach is analogous to the Schwinger treatment of angular momentum by building everything up from doublets of spin $\frac{1}{2}$. One can build up all possible SU_3 multiplets by combining sakaton triplets. This is consistent with the philosophy of the Sakata model which considers that all elementary particles are composites built from the elementary sakaton and antisakaton triplets. This procedure is also relatively simple and is, of course, valid independently of the validity of the Sakata model. This approach is carried out in detail in Appendix A.

Another approach is to use the different SU_2 subgroups of SU_3, noting that within each subgroup the transformations and matrix elements of operators are just those of ordinary angular momentum algebra. We note, for example, that the operators τ_+ and τ_- which move us back and forth horizontally across any multiplet diagram move from one state to another in a given isospin multiplet and the matrix elements are just the usual Clebsch–Gordan coefficients of eq. (2.5). However, instead of defining isospin as transformations of neutrons and protons into one another, leaving the lambda invariant, we could equally well define a different kind of operation which interchanges neutrons and lambdas leaving the proton invariant. Such transformations would move us across the diagram in the direction perpendicular to that of axis 2 in Fig. 3.4. Since these transformations are also two-dimensional unitary transformations, they are described by an angular momentum algebra in which the operators C_+ and B_- play the roles of τ_+ and τ_-. The matrix elements of these operators are again given by ordinary Clebsch–Gordan coefficients of the three-dimensional rotation group. The matrix elements of the operators B_+ and C_- are obtained from a third SU_2 group in which protons and lambdas are transformed into one another, the neutron is left invariant, and we move along lines perpendicular to axis 3 in Fig. 3.4. In dealing with these different SU_2 multiplets we must be careful whenever there is more than one state at a given lattice point as at the origin in Fig. 3.3b. The particular states chosen to be eigenfunctions of T^2 are not the proper states to fit into multiplets which cross the diagrams at angles of 120°. Different linear combinations of these states which are not eigenfunctions of T^2 are necessary to fit into the other SU_2 multiplets. The proper linear combinations, however, are easily determined after some simple algebra. This approach is carried out in detail in Appendix B.

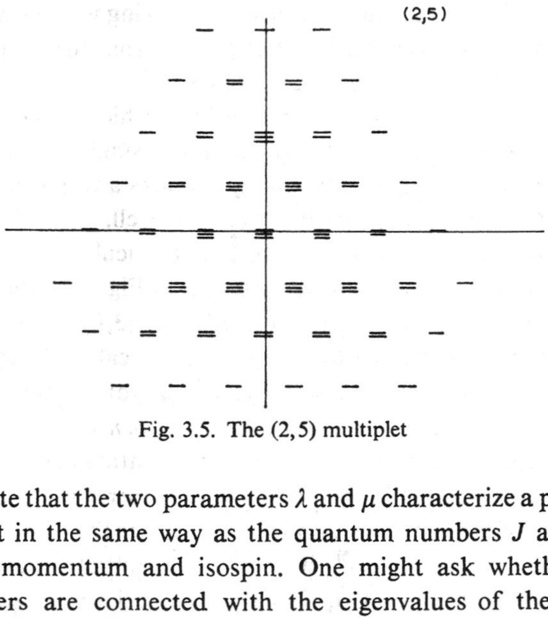

Fig. 3.5. The (2,5) multiplet

We note that the two parameters λ and μ characterize a particular multiplet in the same way as the quantum numbers J and T for angular momentum and isospin. One might ask whether these parameters are connected with the eigenvalues of the Casimir operators C_μ of the SU_3 algebra. This turns out to be the case. However, it is much more convenient to use the numbers λ and μ which have simple integral values to specify the multiplets rather than the eigenvalues of the Casimir operators which turn out to be complicated polynomials in λ and μ. This is analogous to angular momentum where one uses the number J to characterize the multiplet rather than the eigenvalue of the operator J^2 which is the polynomial $J(J+1)$. Further discussion of the Casimir operator is given in Appendix B and in Chapter 4.

Using any of the methods outlined above one finds that there are SU_3 multiplets labeled by the numbers (λ, μ) for all possible integral values of λ and μ. One also finds the following rule for deciding how many states there are at each lattice point:

(a) The outer ring of lattice points is always single with only one state at each lattice point.

(b) Going inward each consecutive ring of lattice points has one more point at each lattice point than the outer ring. This continues until one arrives at a ring which is either a point or a triangle.

(c) Once in going inward one arrives at a ring which is a triangle, the number of states at each lattice point within the triangle is the same as on the perimeter of the triangle.

These rules are illustrated in Fig. 3.5, which shows the (2,5) multiplet. In this multiplet the outer ring is single, the next ring is double, the third ring is triple and this ring is a triangle. Thus, the states within the triangle are all triplets as well.

Some of the multiplets which are of particular interest in elementary particle classification are shown in Fig. 3.6. These are the (3,0), the (0,3) and the (2,2) multiplets. The (3,0) and (0,3) multiplets each have ten states and are sometimes called decuplets.

The operator N, the sum of $\frac{1}{3}$ the baryon number and the strangeness, has eigenvalues having the form n, $n+\frac{1}{3}$ and $n-\frac{1}{3}$ where n is an integer. However, since all of the operators in the set change N by 0 or ± 1 and not by any fractional number, the eigenvalues of N within a given multiplet must all have either the form n, $n+\frac{1}{3}$ or $n-\frac{1}{3}$. Thus there are three different types of multiplets having eigenvalues of N which are either integral, integral$+\frac{1}{3}$, or integral$-\frac{1}{3}$. This characterization of types of multiplets is analogous to the angular momentum multiplets with either integral or half-integral eigenvalues. It can be shown* that the classification of the SU_3 multiplets into these three types is simply expressed in terms of the numbers λ and μ. The quantity $\frac{1}{3}(\lambda-\mu)$ determines the type of

* See Appendix B, eq. (B.18).

multiplet. The eigenvalues of N are integral, integral $+\frac{1}{3}$ or integral $-\frac{1}{3}$ when the value of $\frac{1}{3}(\lambda-\mu)$ is integral, integral $+\frac{1}{3}$ or integral $-\frac{1}{3}$, respectively. This property is illustrated in the simple multiplets of Fig. 3.2, 3.3 and 3.6.

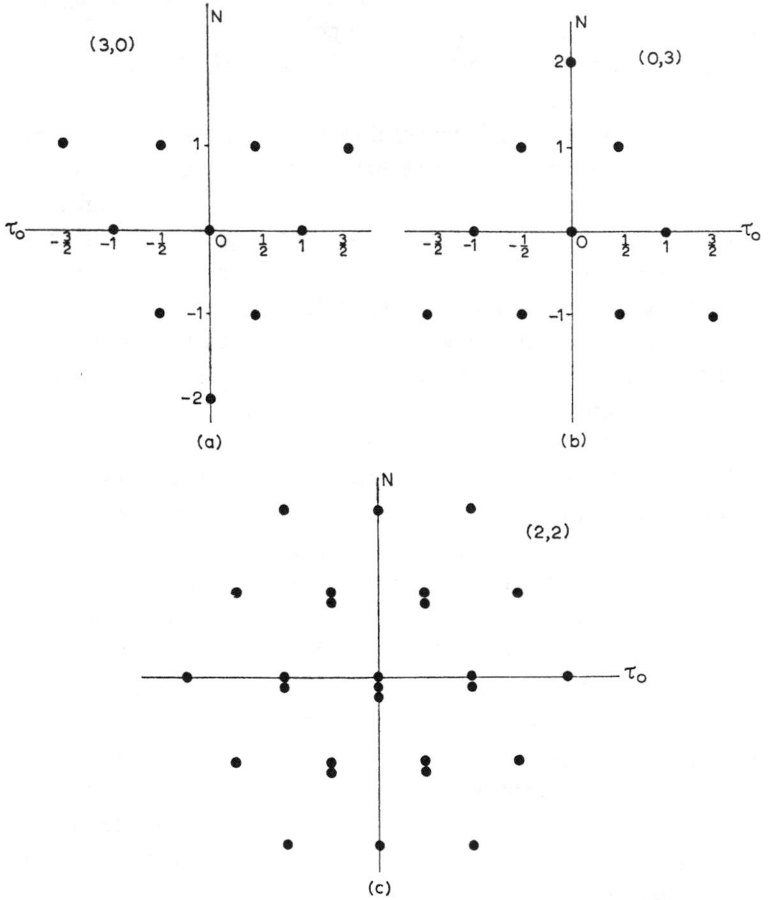

Fig. 3.6

3.3. COMBINING SU_3 MULTIPLETS

In a system consisting of several parts the value of N for the total system is equal to the sum of the values of N for the various parts, since both baryon number and strangeness are additive. One therefore arrives at rules for combining multiplets analogous to the rules for combining integral and half-integral angular momenta.

$$(n) + (n) \quad \rightarrow (n) \tag{3.3a}$$

$$(n \pm \tfrac{1}{3}) + (n) \quad \rightarrow (n \pm \tfrac{1}{3}) \tag{3.3b}$$

$$(n + \tfrac{1}{3}) + (n - \tfrac{1}{3}) \rightarrow (n) \tag{3.3c}$$

More detailed rules for coupling specific multiplets can be obtained by methods analogous to those used for coupling angular momenta. Unfortunately, these are not quite so simple as the angular momentum rules and one cannot remember them as easily as one remembers that coupling $T=5$ to $T=4$ gives all integral values of T from 1 to 9.

One coupling which is of particular interest in elementary particles is the coupling of two octets. For this case, it can be shown* that the 64 states formed from the elements of two octets break up into 6 multiplets.

$$(1,1)+(1,1) \rightarrow \underbrace{(0,0)+(1,1)+(2,2)}_{\mathrm{S}} + \underbrace{(3,0)+(0,3)+(1,1)}_{\mathrm{A}} . \tag{3.4}$$

One finds a singlet, two octets, two decuplets and one 27-uplet. Adding up the total number of states in these multiplets one finds indeed $1+8+27+10+10+8=64$. The two octets might represent two multiplets of the same type, in the same way that two isospin triplets coupled together might represent a two-pion system. For such a case, it is of interest to specify the multiplets of the combined system with regard to their permutation symmetry. For example, in coupling two $T=1$ isospin multiplets the $T=0$ and $T=2$ multiplets for the combined system are symmetric with regard to permutation of the two $T=1$ components whereas the $T=1$ state

* See Appendix A, eq. (A.2).

of the combined system is antisymmetric.

$$(T=1)+(T=1)=\underbrace{(T=0)+(T=2)}_{\text{S}}+\underbrace{(T=1)}_{\text{A}}\,. \qquad (3.5)$$

In the same way one finds that the first three multiplets in eq. (3.4) are the symmetric combinations and the last three are antisymmetric. This is indicated by the letters S and A in eq. (3.4).

One can define a generalized Pauli principle for identical SU_3 multiplets in the same way as for isospin. The wave function for a two-particle system can be written as a product of a space-spin factor and a factor depending on the internal quantum numbers isospin and hypercharge. One then requires the overall wave function to be symmetric for bosons and antisymmetric for fermions, including both the space-spin and internal factors. Thus the states of two bosons both belonging to the same octet must belong to the multiplets labeled S in eq. (3.4) if the space-spin is symmetric and to the multiplets labeled A if the space-spin part is antisymmetric. For fermions the S-multiplets go with antisymmetric space-spin and the A-multiplets with symmetric space-spin.

Note that two (1, 1) octets appear in the coupling of two octets. The analogous situation does not occur in coupling two isospin multiplets, where one never gets more than one isospin multiplet for the combined system having a given value of T. In isospin couplings several multiplets having the same T appear only when one couples at least three isospin multiplets.

An example of such a case is the coupling of two nucleons and a K-meson. Each of these particles is in a $T=\frac{1}{2}$ isospin doublet. Thus there are eight possible states for the two-nucleon–K-system and these can be broken up into three multiplets, one $T=\frac{3}{2}$ quartet and two $T=\frac{1}{2}$ doublets. One can describe the two doublets by saying that in the first doublet the two nucleons are coupled to $T=0$ and the resulting two-nucleon singlet is then coupled to the K-meson to give $T=\frac{1}{2}$. In the second doublet the two nucleons are coupled to $T=1$ and the resulting two-nucleon triplet is then coupled to the K-meson to give $T=\frac{1}{2}$.

$$\begin{aligned} &\{(\mathrm{N}+\mathrm{N})_{T=0}+\mathrm{K}\}_{T=\frac{1}{2}}\,, \\ &\{(\mathrm{N}+\mathrm{N})_{T=1}+\mathrm{K}\}_{T=\frac{1}{2}}\,. \end{aligned} \tag{3.6a}$$

However, this is not the only way to specify the individual $T=\frac{1}{2}$ doublets. Any linear combination of the states of the two doublets is also an isospin doublet and any two orthogonal sets of linear combinations can be used to specify the two doublets. One might, for example, couple one of the nucleons to the K-meson first rather than coupling the two nucleons and in this way specify the two doublets by saying that in one of them a nucleon and the K are coupled to $T=0$ and in the other to $T=1$.

$$\begin{aligned} &\{(\mathrm{N}+\mathrm{K})_{T=0}+\mathrm{N}\}_{T=\frac{1}{2}}\,, \\ &\{(\mathrm{N}+\mathrm{K})_{T=1}+\mathrm{N}\}_{T=\frac{1}{2}}\,. \end{aligned} \tag{3.6b}$$

The two doublets obtained in this way (3.6b) would be linear combinations of the two doublets (3.6a) obtained by coupling the two nucleons first and these two sets of doublets would be related by unitary transformation.

A similar ambiguity exists in the specifications of the two octets arising in eq. (3.4) when coupling together two SU_3 octets. It is possible to distinguish between the two octets in the manner suggested by eq. (3.4), namely, by the permutation symmetry. One of the (1, 1) octets is symmetric with respect to interchange of the two components on the left-hand side of (3.4) and the other is antisymmetric. On the other hand, any linear combination of the symmetric and antisymmetric octets is also an octet although it does not have a definite permutation symmetry. Permutation symmetry may not be important in some physical problems (particularly if the two octets being coupled together represent different kinds of distinguishable particles). In these cases any two orthogonal linear combinations of the two octets on the right-hand side of (3.4) may be chosen to specify the states.

3.4. *R*-SYMMETRY AND CHARGE CONJUGATION

The R-transformation or hypercharge reflection is defined on any state as the reversal of the sign of the quantum numbers N and T_0 of the state accompanied by multiplication by a phase factor determined by convention. This corresponds to an inversion about the origin of the multiplet diagram. For multiplets like the (1, 1) octet which are symmetric about the origin the *R*-transformation carries one state into another state within the same multiplet. For multiplets which are not symmetric about the origin, like the (3,0) decuplet the *R*-transformation carries each state into the corresponding state of the conjugate multiplet; e.g. it carries a member of the (3,0) decuplet into a member of the (0, 3) decuplet. The *R*-transformation is not included in the unitary transformations of the group SU_3. It is thus possible for an interaction to be invariant under SU_3 and not invariant under *R* and vice versa. Experimental evidence seems to indicate that strong interactions are not invariant under the *R*-transformation.

For bosons where particles and their charge conjugates appear in the same SU_3 multiplet, the *R*-transformation is equivalent to charge conjugation. The *R*-transformation is therefore useful in considering properties of boson multiplets. Since boson multiplets include states which are eigenstates of charge conjugation and therefore of the *R*-transformation, a phase convention is necessary to determine the phase of the eigenvalue of *R*. The phase is chosen to be the same as that under charge conjugation; i.e. particles like the π^0 which are even under *C* are even under *R*.

The *R*-transformation is useful in classifying multiplets occurring in the combination of two (1, 1) octets and in distinguishing between the two equivalent octets arising in this combination. If the two octets being combined are both even under *R* (e.g. two pseudoscalar meson octets) some states of the combined system will be even under *R* and others will be odd. The classification of the multiplets under *R* is related to the classification by permutation symmetry but slightly different. The three multiplets arising from states which are symmetric under permutation of the two octets are even under *R*, while the antisymmetric octet is odd under *R*. This is evident from examination of the states in the center of the multiplet diagram with quantum numbers $T_0 = N = 0$. These states are produced by taking linear combinations of states of the two initial octets involving a particle and its antiparticle. Permutation of the members of the two octets is thus equivalent to charge conjugation or to the R-transformation. This argument is not valid for the (3,0) and (0,3) decuplets. Although these contain only states which are antisymmetric under permutations they do not contain any states involving a particle and its corresponding antiparticle. Charge conjugation or the R-transformation on any state in the (3,0) decuplet lead to the corresponding state in the (0,3) decuplet. The states in the two decuplets are therefore not eigenstates of the R-transformation.

In considering reactions, decays or couplings to two-boson states, the requirement of invariance under charge conjugation reduces the number of channels. In particular, charge conjugation invariance removes the ambiguity of two octets arising in the coupling of two octets. If the two octets being coupled are equivalent bosons, one of the octets arising in the combined system is even under C, while the other is odd.

3.5. THE GENERALIZATION TO ANY SU_3 ALGEBRA

We have used the Sakata model for elementary particles to develop the SU_3 Lie algebra and determine the structure of the multiplets. However, the algebra of SU_3 does not depend on the Sakata model; the latter is merely a convenient way to introduce the algebra. This is analogous to building up angular momentum or SU_2 algebra by using the basic spin one-half objects, e.g. nucleons.

If a set of eight operators satisfying commutation rules like those of the operators (3.2) should arise in any physical problem we now know that these operators constitute the Lie algebra of the group SU_3. These operators can be used to characterize states of the associated quantum mechanical system and group them into multiplets. These multiplets will have the same structure that we have found using the Sakata model for elementary particles, since multiplet structure depends only on the Lie algebra and not on the particular model used. The only possible difference between multiplet structure determined by a particular model and as determined from the Lie algebra is the possibility that some multiplets which are possible in the Lie algebra may be absent from the particular model. For example, if orbital angular momentum had been used to build up the structure of angular momentum multiplets the possibility of half-integral eigenvalues for J would have been missed. We have no proof that the Sakata model gives all the possible SU_3 multiplets. This happens to be the case but it will not be proved explicitly in this book.

The Sakata model for elementary particles does not seem to be in agreement with experiment at this time.

3.6. THE OCTET MODEL OF ELEMENTARY PARTICLES

Let us now examine the classification of states of elementary particles from a somewhat different point of view. The experimentally found elementary particles and resonances can be grouped into sets of states all having the same spin and parity but differing by 'internal' quantum numbers such as charge and strangeness. One might hope to classify these sets of particles into multiplets corresponding to some Lie algebra. The operators of this Lie algebra would then change only the internal quantum numbers of the state and would not affect spin, parity, or any of the spacial variables. Particles having the same spin, parity and strangeness but different electric charge are grouped into isospin multiplets. We are therefore looking for a higher symmetry, in which the multiplets would include several isospin multiplets having different values of strangeness. The electric charge and the strangeness are two additive conserved quantities which can be used to specify the internal quantum numbers of a particle, and there is no other quantity of this kind in evidence. This suggests that the Lie algebra desired is one of rank two. The baryon number is also a conserved additive quantum number but is not relevant to this discussion since there does not yet seem to be any physical interest in placing particles having different values of the baryon number in the same multiplet.

Let us now examine the experimentally observed sets of particles to see whether they can be grouped into multiplets in a natural way. We first examine two-dimensional plots of all known particles of a given spin and parity. The coordinate axes might be electric charge and strangeness since these are the two quantities conserved. However, since we know that we wish to include isospin in the higher symmetry a plot of T_z and the hypercharge Y are more suitable variables. This can be seen from Fig. 3.7, which shows plots for all the known stable baryons of spin $\frac{1}{2}$, all the known pseudoscalar mesons and all the known vector meson resonances. These diagrams immediately suggest the octet multiplets for the group SU_3 with an additional singlet vector meson. The meson octets are just like those predicted by the Sakata model. However, it appears natural from the plot of Fig. 3.7 to place the baryons also in an octet rather than having some of them in a triplet as in the Sakata model. This classification of elementary particles is called the octet model of unitary symmetry or the eightfold way.

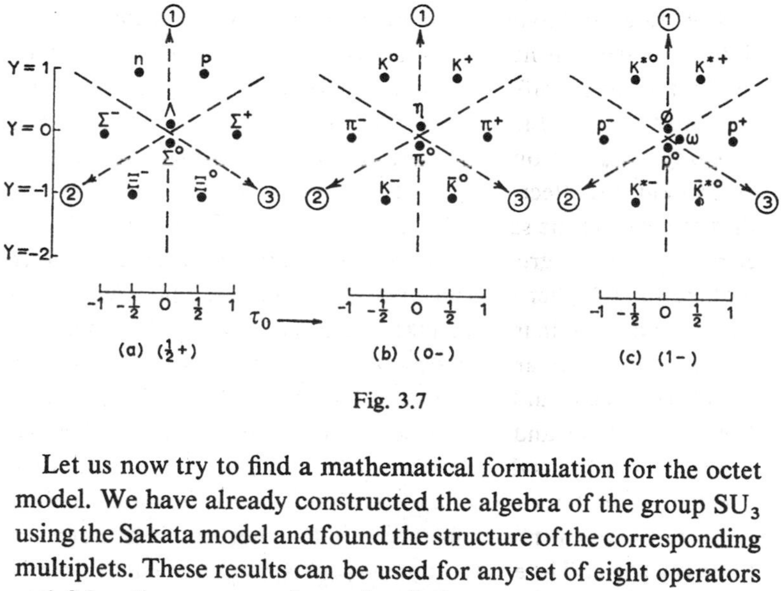

Fig. 3.7

Let us now try to find a mathematical formulation for the octet model. We have already constructed the algebra of the group SU_3 using the Sakata model and found the structure of the corresponding multiplets. These results can be used for any set of eight operators satisfying the commutation rules (3.2) even though they have no connection with the Sakata model (this is analogous to the application of all angular momentum results to isospin, even though isospin has no connection with a physical three-dimensional rotation).

We define for the octet model eight operators satisfying the commutation rules (3.2). The three isospin operators τ_+, τ_- and τ_0 are defined in the conventional way. From Fig. 3.7, particularly from the baryon octet, we see that the diagonal operator N should correspond to the hypercharge $Y=B+S$ rather than to $\frac{1}{3}B+S$ as in the Sakata model. Once the four operators τ_+, τ_-, τ_0 and N are defined, the remaining four operators are completely defined by the commutation relations (3.2). The relations (3.2a) establish them as step operators changing the eigenvalue of τ_0 by $\pm\frac{1}{2}$ and the eigenvalue of N by ± 1. The relations (3.2b) give the commutators of the step operators among themselves. Using these commutation rules and the results of multiplet structure obtained from the Sakata model the matrix elements of any one of these eight operators can be calculated between any two states of a given multiplet. We therefore have a complete specification of these operators.

Since the hypercharge Y has only integral values only states with integral eigenvalues of N can occur in the octet model. No multiplets with third-integral eigenvalues of N can occur. Thus only those multiplets (λ, μ) occur for which $\frac{1}{3}(\lambda-\mu)$ is an integer. The simplest multiplets occurring in the octet model are thus the (0,0) singlet, the (1,1) octet, the (3,0) and (0,3) decuplets and the 27-dimensional (2,2).

There is no simple description of the physical basis of the octet model in terms of interactions between particles, analogous to the charge independence of nuclear forces for isospin symmetry and the equivalence of neutron–proton and lambda interactions for the Sakata model. In the Sakata model the members of multiplets more complicated than the sakaton triplets (e.g. the meson octets) are not all equivalent and have different interactions. The unitary symmetry requires merely that these interactions be related in such a way as to preserve the equivalence of the interactions between the three basic sakatons. This is analogous to the requirement by isospin symmetry that the forces between Σ's and π's have relations between them which preserve the charge independence of nuclear forces. In the octet model the eight basic baryons are not equivalent, and the interactions between different pairs of baryons are related but not identical. This is analogous to the hypothetical case discussed in isospin where one considered the interactions between Σ's and π's resulting from isospin symmetry in the case where nucleons did not exist. The interactions between the mesons and baryons in the two (1, 1) octets are related as if there existed some fictitious triplet like the sakaton for which the forces were *really* independent. However, the existence of a basic sakaton-type triplet is not necessary for the application of SU_3 symmetry, just like the existence of particles of half-integral spin is not necessary in order to allow one to use angular momentum. A basic triplet for the octet model would need to have very peculiar properties since it would have third-integral hypercharge. Hypothetical triplets have been used in mathematical presentations of the octet model requiring both a boson triplet and fermion triplet, and there has also been a suggestion that a triplet with third-integral electric charge might exist.

There is no simple definition for SU_3 operators in the octet model analogous to the definition (3.1) for the Sakata model. One can define operators for the octet model in terms of creation and annihilation operators but these definitions are rather cumbersome. The procedure is directly analogous to the extension of the definition of isospin operators (2.1) to include pions and hyperons as well as nucleons. One would have to add terms involving creation and annihilation operators for all these particles. Each term would have a numerical coefficient differing from unity for particles like pions which are not members of an isospin doublet and determined by the relations (2.5). However, as soon as the simplicity of the relations (2.1) and (3.1) is lost one may just as well use relations of the form (3.5) giving matrix elements of the operators within any multiplet to define the operators rather than an explicit definition in terms of creation and annihilation operators.

Let us now examine the kind of experimental predictions which can be made on the basis of the octet model of SU_3. By analogy with isospin we can make the following observations.

A. Predictions from the multiplet structure

1. *Classification.* All new resonances or strongly interacting particles which are found must belong to SU_3 multiplets. Once one member of a given multiplet is found, all the other members of the multiplet must also exist.

2. *Couplings.* If one considers resonances between two particles, the multiplet structures which can arise are determined by the SU_3 coupling rules. Thus, all resonances of nucleons and hyperons with π- and K-mesons must belong to the multiplets obtained by combining two (1, 1) octets; namely, the (0, 0) singlet, the (1, 1) octet, the (2, 2) with 27 states, the (3, 0) decuplet and the (0, 3) decuplet.

B. Predictions of relations between matrix elements

1. *Decays.* The decay rates or widths of different resonances belonging to the same multiplet are related by SU_3 coupling rules involving coefficients analogous to Clebsch–Gordan coefficients.

2. *Reactions.* Cross sections for different reactions involving members of the same multiplet are related by SU_3 coupling rules.

3. *Selection rules.* One may find selection rules which result from SU_3 couplings.

C. Symmetry-breaking effects

Relations may be found for processes involving interactions like the electromagnetic interaction which are not invariant under SU_3 but have simple transformation properties.

A detailed analysis of the possible experimental predictions from the octet model is given in Appendix C.

3.7. THE MOST GENERAL SU_3 CLASSIFICATION

Let us now attempt to apply the group SU_3 to the classification of elementary particles in the most general manner possible, without specifically assuming the Sakata or octet model. We begin with the eight operators (3.1) defined now in terms of their commutation rules rather than the specific representation in the Sakata model. We also identify the τ-operators with isospin and the operator N with strangeness. However, we no longer require that $N = \frac{1}{3}B + S$ or $N = B + S$; rather we simply require that changes of N within a multiplet be equivalent to changes of strangeness. The multiplets still have the same structure as determined from the Sakata model and the classification of the multiplets according to the eigenvalues of N as being integral, integral $+\frac{1}{3}$, and integral $-\frac{1}{3}$ is still valid.

These can be obtained from the algebra of the operators without assuming the Sakata model. The relations (3.3) for combining multiplets are also still valid.

Let us now consider what kind of multiplets can be used for the nucleons and the pions. The pions are bosons and have no selection rule against their being created singly without changing or creating other particles. Nucleons, for example, can simply emit an arbitrary number of pions and remain nucleons. From this property of the pions and the relations (3.3) we see that the pions must be in a multiplet having an integral value of N. The nucleons are fermions and because of baryon number conservation can only be produced in pairs with antibaryons. The nucleons can therefore be put either into a multiplet having an integral value of N or a third-integral. If the nucleons are put into multiplets having integral values of N then the antinucleons must also be in multiplets having integral values of N. If the nucleons are put into multiplets having values of N equal to an integral $+\frac{1}{3}$, the antinucleons must be put into multiplets in which the eigenvalues of N are integral $-\frac{1}{3}$. Because of the symmetry between particles and antiparticles, there is no physical difference between a scheme where the nucleons are in $n+\frac{1}{3}$ multiplets and the antinucleons are in $n-\frac{1}{3}$ multiplets and vice versa. There are therefore two possibilities. Either the nucleons and antinucleons are in third-integral multiplets and the mesons are in integral multiplets, or both the nucleons and the pions are in integral multiplets.

If the nucleons are put into the simplest third-integral multiplets, we obtain the Sakata model which puts the nucleons into a (1,0) multiplet and leads to the relation (3.1) that the quantum number N is just $\frac{1}{3}B+S$. One might look for other possible multiplets with third-integral N as a different choice for the nucleons. However, there does not seem to be any other reasonable classification which fits with the known particles.

If both the nucleons and mesons are put into integral multiplets, the smallest possible multiplet with integral N (excluding the singlet) is the (1, 1) octet. The only feasible arrangement for nucleons and pions in these multiplets is to put both the nucleons and the pions each into an octet as shown in Fig. 3.7. This is just the octet model. In this model only integral values of N occur and N is the hypercharge $Y=B+S$. At the time of the writing of this book, the octet model looks hopeful for a description of elementary particles and the Sakata model has been discarded because of its disagreement with experiment.

Note that in both Sakata and octet models, the mesons are in integral multiplets, while the baryons are in third-integral in the Sakata and integral in the octet model. The classification of bosons can thus be the same in both models, but the classification of baryons cannot. Note that if bosons should be found which are classified into *third-integral* multiplets, their decay into any combination of nucleons, antinucleons and pions would be forbidden in both the Sakata and octet models. The same would be true for baryons in integral multiplets in the Sakata model or in third-integral multiplets in the octet model. Such particles would be analogous to the strange particles in the isospin classification whose decay into nucleons and pions is forbidden by isospin.

CHAPTER 4

THE THREE-DIMENSIONAL HARMONIC OSCILLATOR

4.1. THE QUASISPIN CLASSIFICATION

The energy levels of a three-dimensional harmonic oscillator are known to be highly degenerate. One way of seeing this is to note that the harmonic oscillator Hamiltonian separates in Cartesian coordinates into three independent oscillators all having the same frequency. The energy of the oscillator then depends on the total number of oscillator quanta present and is independent of the distribution of the quanta between the three oscillators.

The three-dimensional harmonic oscillator is also soluble in spherical coordinates and has the characteristic degeneracy of rotational invariance. However, there is additional degeneracy beyond that of rotational invariance. There are also degenerate states corresponding to different eigenvalues of the total angular momentum.

Let us consider the possibility of describing this degeneracy in terms of operators which, when acting on one state of the oscillator, give another degenerate state. Such operators have matrix elements only between degenerate states and must commute with the Hamiltonian. Commutators of such operators would also commute with the Hamiltonian and therefore a Lie algebra should be generated by these operators. Since we know that angular momentum operators must commute with the Hamiltonian because of its rotational invariance, we should expect to find a Lie algebra greater than that of angular momentum and including the angular momentum Lie algebra.

The harmonic oscillator Hamiltonian is

$$H = \frac{1}{2m}(p_1^2+p_2^2+p_3^2) + \frac{m\omega^2}{2}(x_1^2+x_2^2+x_3^2)\,, \qquad (4.1)$$

where m is the mass of the oscillator and ω is the frequency. The operators which do not change the energy of a state are clearly those which reduce the number of oscillator quanta in one direction and increase the number in another direction, thereby keeping the total number of quanta constant. These are most conveniently expressed in terms of the oscillator creation and annihilation operators. We therefore define

$$a_\mu = \left(\frac{m\omega}{2\hbar}\right)^{\frac{1}{2}} x_\mu + \frac{\mathrm{i}}{(2m\omega\hbar)^{\frac{1}{2}}} p_\mu\,, \qquad \mu = 1,2,3\,;$$

$$a_\mu^\dagger = \left(\frac{m\omega}{2\hbar}\right)^{\frac{1}{2}} x_\mu - \frac{\mathrm{i}}{(2m\omega\hbar)^{\frac{1}{2}}} p_\mu\,, \qquad \mu = 1,2,3\,. \qquad (4.2)$$

These operators satisfy the boson commutation relations

$$[a_\mu, a_\nu^\dagger] = \delta_{\mu\nu}\,,$$

$$[a_\mu^\dagger, a_\nu^\dagger] = [a_\mu, a_\nu] = 0\,. \qquad (4.3)$$

When expressed in terms of these operators, the Hamiltonian (4.1) assumes the simple form

$$H = \tfrac{1}{2}\hbar\omega \sum_{\mu=1}^{3} (a_\mu^\dagger a_\mu + a_\mu a_\mu^\dagger) = \hbar\omega\left(\sum_{\mu=1}^{3} a_\mu^\dagger a_\mu + \tfrac{3}{2}\right). \qquad (4.4)$$

The angular momentum operators are given by

$$l_{\mu\nu} = x_\mu p_\nu - p_\mu x_\nu = \mathrm{i}(a_\mu a_\nu^\dagger - a_\nu a_\mu^\dagger)\,. \qquad (4.5)$$

Operators having the form $a_\mu^\dagger a_\nu$ clearly commute with the Hamiltonian and connect only degenerate states of the oscillator, since they transfer a quantum from the ν-direction to the μ-direction. As there are three possible values for μ and ν, there are nine such operators. These look very similar to the operators of the group SU_3, eq. (3.1) in the Sakata model of elementary particles. Although the proton, neutron and Λ-operators are fermion creation and annihilation operators whereas the harmonic oscillator operators are boson operators, the commutation relations of the bilinear products are the same. Again we can find one operator of the set of nine which commutes with all the rest. It is just the Hamiltonian operator (4.4) directly **analogous** to the baryon number B in the case of the Sakata model.

Operators analogous to those of the Sakata model can be written very simply by replacing the sakaton creation and annihilation operators by corresponding harmonic oscillator operators (4.2). However, the treatment of the Sakata model made use of an SU_2 subgroup, namely isospin, by picking a preferred direction in the npΛ-space; namely the Λ-direction. The analogous procedure would be to pick the x_3-direction as a preferred direction for the oscillator and choose the SU_2 subgroup to be that of the two-dimensional harmonic oscillator in the space defined by the coordinates x_1 and x_2. For this purpose it is convenient to define boson operators for cylindrical coordinates

$$\begin{aligned} a_{\pm} &= (a_x \mp i a_y)/\sqrt{2}\,, \\ a_0 &= a_3\,, \\ a_{\pm}^{\dagger} &= (a_x^{\dagger} \pm i a_y^{\dagger})/\sqrt{2}\,, \\ a_0^{\dagger} &= a_3^{\dagger}\,. \end{aligned} \tag{4.6}$$

In terms of these operators we can write

$$H = \hbar\omega\{a_+^{\dagger} a_+ + a_-^{\dagger} a_- + a_0^{\dagger} a_0 + \tfrac{3}{2}\}\,, \tag{4.7a}$$

$$\lambda_+ \equiv a_+^{\dagger} a_-\,, \tag{4.7b}$$

$$\lambda_- \equiv a_-^{\dagger} a_+\,, \tag{4.7c}$$

$$\lambda_0 \equiv \tfrac{1}{2}(a_+^{\dagger} a_+ - a_-^{\dagger} a_-) = \tfrac{1}{2} l_3\,. \tag{4.7d}$$

The operators a_+ and a_- are now analogous to the proton and neutron operators and the operators λ_+, λ_- and λ_0 are analogous to the isospin operators. However, the λ-operators are not the angular momentum operators for the harmonic oscillator. They are really the operators for the group SU_2 of the two-dimensional harmonic oscillator and not those of the three-dimensional rotation group.

Note that the operator λ_0 is proportional to the 12-component of the angular momentum but multiplied by a factor of $\frac{1}{2}$. Although only integral values of the orbital angular momentum l can occur in the harmonic oscillator, the quantum number λ_0 can have either integral or half-integral eigenvalues, depending upon whether l is even or odd. The three λ-operators can be called quasispin operators since they satisfy angular momentum commutation rules. By analogy with isospin one can define the total quasispin operator

$$\lambda^2 = \tfrac{1}{2}\{\lambda_+ \lambda_- + \lambda_- \lambda_+\} + \lambda_0^2 \,. \tag{4.8}$$

In addition to the quasispin operators, one can define the remaining operators of the algebra by direct analogy with the corresponding operators of the Sakata model,

$$\begin{aligned} B_+ &= a_+^\dagger a_0 \,, & B_- &= a_-^\dagger a_0 \,, \\ C_+ &= a_0^\dagger a_- \,, & C_- &= a_0^\dagger a_+ \,, \end{aligned} \tag{4.9}$$
$$N = \tfrac{1}{3}(a_+^\dagger a_+ + a_-^\dagger a_- - 2a_0^\dagger a_0) = \tfrac{1}{3}(a_1^\dagger a_1 + a_2^\dagger a_2 - 2a_3^\dagger a_3) \,.$$

Now that we have defined operators satisfying exactly the same commutation rules as those of the Sakata model, we can use all of the results for the group SU_3 obtained from the Sakata model to describe the multiplets of degenerate eigenstates of the harmonic oscillator Hamiltonian. We can plot multiplet diagrams in which the eigenvalue of N is plotted against the eigenvalue of λ_0. The quantum number N is one-third the difference between the number of quanta in the 12-plane and twice the number of quanta in the 3-direction. $N=0$ when the numbers of quanta in all three directions are equal, $N>0$ when the number of quanta in the 3-direction is less than the average number in the 1- and 2-directions and $N<0$ when the number of quanta in the 3-direction is greater than the average in the 1- and 2-directions. The quantum number N therefore measures a 'deformation' or departure from spherical symmetry.

The ground state of the harmonic oscillator is non-degenerate and clearly is a (0,0) singlet of SU_3. The first excited level of the oscillator has a three-fold degeneracy corresponding to a single oscillator quantum which can be in any of the three directions. This is just the (1,0) triplet of SU_3 corresponding to the sakaton. The expected third-integral eigenvalues of N occur. Since the oscillator quantum corresponds to the sakaton, the nth excited level of the oscillator containing n oscillator quanta corresponds to a state of n sakatons. Furthermore, since the oscillator quanta are bosons, a state of n oscillator quanta must be totally symmetric with respect to permutation of the quanta. The nth excited level of the harmonic oscillator must therefore correspond to the totally symmetric SU_3 multiplet obtained from the n-sakaton system. This is shown in Appendix A to be the $(n,0)$ triangular multiplet.

The levels of the nth excited state can be classified into quasispin multiplets having all possible values of the total quasispin λ from 0 to $\frac{1}{2}n$, including both integral and half-integral values. The total number of states in this $(n,0)$ multiplet is $\frac{1}{2}(n+1)(n+2)$. This is just the number of states that one obtains by examining all possible ways of distributing n oscillator quanta among the three oscillator directions. The states of the $(n,0)$ multiplet thus completely exhaust the degeneracy of the nth level of the harmonic oscillator. The degeneracy of the three-dimensional harmonic oscillator is thus completely described by the classification of its states using the group SU_3.

4.2. THE ANGULAR MOMENTUM CLASSIFICATION

The quasispin classification of the levels of the three-dimensional harmonic oscillator is not the conventional one and does not display explicitly the symmetry of the oscillator under ordinary three-dimensional rotations. One normally classifies the state of the three-dimensional oscillator by expressing the wave functions either in Cartesian coordinates or spherical coordinates. In the latter case they are classified by taking eigenfunctions of the orbital angular momentum l. We know that if n is even, all even values of l occur up to $l=n$, whereas if n is odd, all odd values of l occur up to $l=n$. The states in the quasispin classification are naturally related to those in the angular momentum classification. In both cases l_3 is a good quantum number. The quasispin multiplets corresponding to integral values of λ have only even values of l_3; those corresponding to half-integral eigenvalues of λ have only odd values of l_3. Let us examine the quasispin multiplets in the $(n,0)$ multiplet corresponding to the nth excited oscillator level. The two largest quasispin multiplets corresponding to $\lambda=\frac{1}{2}n$ and $\lambda=\frac{1}{2}(n-1)$ just contain all the eigenvalues of l_3 appearing in the $l=n$ angular momentum multiplet. Continuing in this manner, we find the expected one-to-one correspondence between the eigenvalues of l_3 arising in the quasispin and the angular momentum classification. However, for all values of l_3 where more than one state appears in the $(n,0)$ multiplet, those states which are eigenfunctions of the quasispin λ are not eigenfunctions of the orbital angular momentum l^2 and vice versa, since the operators λ^2 and l^2 do not commute.

The classification of states of the harmonic oscillator using the group SU_3 but using eigenfunctions of the orbital angular momentum l is more difficult than the classification using the quasispin operators λ. There is no linear combination of the operators of the Lie algebra (4.9) which commutes with the orbital angular momentum operators, analogous to the operator N which commutes with all the quasispin operators. There therefore is no simple way to represent the multiplets on a diagram analogous to those used for elementary particles because there is no quantum number analogous to N to label the vertical axis. There is no simple way of defining another quantum number in addition to l^2 and l_3 to classify the states in an SU_3 multiplet. The additional quantum number is not necessary to label the states of a single three-dimensional oscillator where the quantum numbers l^2 and l_3 are sufficient to classify the states in any of the triangular $(n,0)$ multiplets. If the quasispin classification is used, the quasispin quantum numbers λ^2 and λ_0 are already sufficient for the classification and the additional quantum number N is redundant, being determined uniquely by the values of λ^2 and λ_0. This is no longer true in SU_3 multiplets which are not triangular where several states having the same quasispin occur with different values of N. Such multiplets arise in the classification of the states of a system of several three-dimensional harmonic oscillators. In such a case, the classification of the states using angular momentum rather than quasispin is difficult because several states arise having the same angular momentum quantum numbers. The additional quantum number needed to distinguish between them is not easily defined.

We thus see one essential difference between the SU_2 subgroup of the group SU_3 and the R_3 or three-dimensional rotation subgroup of SU_3. Although both of these groups have a Lie algebra consisting of the three operators satisfying angular momentum commutation rules, the geometrical and physical significance of the two groups is quite different. The SU_2 subgroup consists of unitary transformations in a two-dimensional space which is a subspace of the three-dimensional space in which the SU_3 transformations are defined. The R_3 group is a group of real (not complex) rotations in the whole three-dimensional space defined by the group SU_3.

In using the angular momentum classification of harmonic oscillator states, a different set of linear combinations of the eight operators (4.9) of the SU_3 Lie algebra is more convenient.

$$l_0 = (a_+^\dagger a_+ - a_-^\dagger a_-) \qquad = 2\lambda_0 , \tag{4.10a}$$

$$l_\pm = \pm\sqrt{2}(a_0^\dagger a_\mp - a_\mp^\dagger a_0) \qquad = \pm\sqrt{2}(C_\pm - B_\pm), \tag{4.10b}$$

$$q_{\pm 2} = -\sqrt{6}a_\pm^\dagger a_\mp \qquad = -\sqrt{6}\lambda_\pm , \tag{4.10c}$$

$$q_{\pm 1} = \mp\sqrt{3}(a_0^\dagger a_\mp + a_\pm^\dagger a_0) \qquad = \mp\sqrt{3}(C_\pm + B_\pm) , \tag{4.10d}$$

$$q_0 = 2a_0^\dagger a_0 - a_+^\dagger a_+ - a_-^\dagger a_- = -3N . \tag{4.10e}$$

The eight operators now appear as the three orbital angular momentum operators and a set of five operators which transform under rotations like the elements of a second rank tensor. The latter are in fact just a linear combination of the quadrupole moment tensors in configuration and momentum space. The operator q_0 is proportional to N but commonly normalized to have integral rather than third-integral eigenvalues. The negative sign is chosen to give it the conventional sign of a quadrupole moment. Since no linear combination of the five quadrupole operators commutes with all the angular momentum operators, we see that there is indeed no operator analogous to N in the quasispin representation which can be simultaneously diagonalized along with l^2 and l_0. The commutation rules of the operators (4.10) are

$$[l_0, l_\pm] = \pm l_\pm\,, \tag{4.11a}$$

$$[l_+, l_-] = 2l_0\,, \tag{4.11b}$$

$$[l_0, q_m] = mq_m\,, \tag{4.11c}$$

$$[l_\pm, q_m] = \sqrt{6-m(m\pm 1)}\,q_{m\pm 1}\,, \tag{4.11d}$$

$$[q_0, q_{\pm 1}] = \pm\tfrac{3}{2}\sqrt{6}\,l_\pm\,, \tag{4.11e}$$

$$[q_1, q_{-1}] = -3l_0\,, \tag{4.11f}$$

$$[q_2, q_{-2}] = 6l_0\,, \tag{4.11g}$$

$$[q_{\pm 2}, q_{\mp 1}] = \pm 3l_\pm\,, \tag{4.11h}$$

$$[q_0, q_{\pm 2}] = [q_{\pm 1}\,, q_{\pm 2}] = 0\,. \tag{4.11i}$$

Let us now attempt to find one of the Casimir operators which commutes with all of the eight operators. We first look for an operator which is a quadratic form in the operators (4.10) by analogy with the operator l^2 for angular momentum. Since the Casimir operator must commute with all the operators (4.10) it must commute with the angular momentum operators and therefore be a scalar under rotations. There are only two scalars which can be constructed that are quadratic in these operators, namely the total angular momentum l^2 and the square of the quadrupole moment. Examining the commutators of an arbitrary linear combination of these two operators with q_m, we find one particular linear combination which commutes with all of them.

$$C = \tfrac{1}{36}[3l^2 + \sum_m (-)^m q_m q_{-m}] \tag{4.12a}$$

$$= \tfrac{1}{36}\{9N^2 + 12\lambda_0^2 + 6(\lambda_+ \lambda_- + \lambda_- \lambda_+ + B_+ C_- + C_- B_+ + B_- C_+ + C_+ B_-)\}, \tag{4.12b}$$

$$[C, q_m] = [C, l_m] = 0, \tag{4.12c}$$

where the coefficient $\frac{1}{36}$ is a conventional normalization factor.

4.3. SYTEMS OF SEVERAL HARMONIC OSCILLATORS

The algebra of the group SU_3 can also be used to classify the states of a system of several harmonic oscillators. An example of such a system is the harmonic oscillator nuclear shell model in which a number of particles are assumed to move independently in a harmonic oscillator potential. The treatment for a single oscillator is easily extended to this case by defining corresponding oscillators for each particle and defining the operators of the Lie algebra by summing over all particles. Let $x_{\mu i}$ and $p_{\mu i}$ be the coordinates and the momenta of the ith particle and $a_{\mu i}$ and $a^\dagger_{\mu i}$ be the corresponding annihilation and creation operators (4.2). We can define the quasi-spin λ_i, the angular momentum l_i and the quadrupole tensor q_i for the ith particle. Operators for a Lie algebra describing the whole system can be defined by summing eq. (4.10) over all of the particles.

$$L = \sum_i l_i, \tag{4.13a}$$

$$\Lambda = \sum_i \lambda_i, \tag{4.13b}$$

$$Q_m = \sum_i q_{im}, \tag{4.13c}$$

$$C = \tfrac{1}{36}[3L^2 + \sum_m (-)^m Q_m Q_{-m}]. \tag{4.13d}$$

We assume that the operators B_+, B_-, C_+, C_- and N are now defined as sums over all particles. The summing over all particles does not affect the commutation relations of the operators and we again have an SU_3 Lie algebra.

The states of the system can be classified according to the multiplets of SU_3 by noting that each individual particle is in a state corresponding to a particular SU_3 multiplet and combining these multiplets according to the rules for coupling SU_3 multiplets. Let us first consider a system of several particles each in the first excited energy level of the oscillator potential (the p-shell). This corresponds in the nuclear shell model to those nuclei having the lowest oscillator shell filled and a number of particles in the next shell. There is a high degeneracy of states for this system, as each particle has a three-fold degeneracy (since nucleons are fermions and cannot occupy the same state, the degeneracy is reduced somewhat in the nuclear model). We can classify the degenerate states of the system by noting that the three states for a single particle are just the sakaton triplet (1,0). The multiplets which arise in the system of n particles in the p-shell are just those arising in the n-sakaton system subject to the restrictions of permutation symmetry including the spins and isospins of the particles.

The second excited level of the oscillator has six states and corresponds to the (2,0) SU_3 multiplet. Nuclei having the lowest two shells filled and the second excited level (the sd-shell) partially filled, can be treated by coupling together the appropriate number of (2,0) multiplets. Note that the closed shells do not contribute to the classification since a closed shell always corresponds to a singlet (0,0) multiplet.

One can now ask what useful purpose is served by classifying states of the harmonic oscillator using the group SU_3. For the single oscillator, we see that the group may give some insight into the additional degeneracy of the system beyond that of angular momentum. However, all the states of the oscillator are conveniently classified by the principal quantum number and the angular momentum quantum numbers and no further quantum numbers are needed. For the case of many particles in an oscillator potential it is possible to find several states having the same angular momentum quantum numbers. Additional quantum numbers are then necessary to specify the states completely and the SU_3 classification may be useful.

4.4. THE ELLIOTT MODEL

A case of particular interest is the model first proposed by Elliott in which an additional two-body interaction of the quadrupole type is added to the oscillator Hamiltonian to remove some of the degeneracy. The Hamiltonian has the form

$$H = H_{\text{osc}} - V \sum_{ijm} (-)^m q_{mi} q_{-mj} \tag{4.14}$$

where H_{osc} represents the Hamiltonian for the particles moving independently in the harmonic oscillator well. The remaining term is a two-body interaction which might be considered as the quadrupole term in the expansion of a general two-body interaction in spherical harmonics. The Hamiltonian (4.14) can be rewritten

$$\begin{aligned} H &= H_{\text{osc}} - V \sum_{m} (-)^m Q_m Q_{-m} \\ &= H_{\text{osc}} - 36CV + 3VL^2 \,. \end{aligned} \tag{4.15}$$

From eq. (4.15) we see that the quadrupole interaction is just the sum of a term proportional to the Casimir operator C and a term proportional to the total orbital angular momentum L^2. The eigenfunctions of the Hamiltonian with the quadrupole interaction are thus those linear combinations of the degenerate harmonic oscillator functions which are simultaneous eigenfunctions of C and L^2. Since any state which is a member of an SU_3 multiplet is automatically an eigenstate of the Casimir operator C, the required eigenfunctions are obtained by first classifying the states of the system into SU_3 multiplets and then choosing the states within the multiplets to be eigenfunctions of the total angular momentum L^2. The energy spectrum exhibits a splitting between the SU_3 multiplets which is proportional to the eigenvalue of the Casimir operator C and a splitting within the multiplets which is proportional to the square of the angular momentum. Since an energy spectrum proportional to L^2 is just a rotational spectrum, we see that each SU_3 multiplet constitutes a rotational band whose 'moment of inertia' is given by

$$\frac{1}{2\mathscr{I}} = 3V\,. \tag{4.16}$$

The SU_3 model was first used to demonstrate how 'collective' features, such as rotational bands, could be obtained from an independent-particle shell-model treatment. The SU_3 classification is useful in shell-model calculations where the residual interaction is reasonably well represented by a quadrupole force and the remaining portion of the residual interaction can be treated as a perturbation. Note that this particular case is one in which the Hamiltonian does not commute with all of the operators of the Lie algebra but commutes with the Casimir operator and the angular momentum operators. Thus the SU_3 quantum numbers can still be used to classify the states, and the members of a given multiplet are not degenerate eigenfunctions of H but have a simple energy spectrum determined by the algebra of the operators; in this case simply proportional to L^2.

Because the Hamiltonian does not commute with all the operators of the Lie algebra, we are not at liberty to choose the particular members of the algebra we wish to be diagonal. We are compelled to take L^2 and one of the angular momentum operators and cannot use the more convenient quasispin operators Λ. Elaborate mathematical techniques have been developed for using the representation in which Λ and N are diagonal and then projecting out states having a good angular momentum. These techniques are beyond the scope of this treatment.

CHAPTER 5

ALGEBRAS OF OPERATORS WHICH CHANGE THE NUMBER OF PARTICLES

5.1. PAIRING QUASISPINS

The isospin Lie algebra was obtained by considering all bilinear products of neutron and proton creation and annihilation operators which do not change the number of particles. Let us now include those bilinear products which change the number of particles. There are just two independent operators

$$s_+ = a_p^\dagger a_n^\dagger \,, \tag{5.1a}$$

$$s_- = a_n a_p \,. \tag{5.1b}$$

These operators create and annihilate respectively a neutron–proton pair in the same quantum state. These two operators together with the three isospin operators and the baryon number (2.1) constitute the six linearly independent bilinear products which can be made from the proton and neutron operators for a single quantum state. Note that if nucleons were bosons rather than fermions, there would be four additional operators corresponding to the creation and annihilation of neutron and proton pairs having both particles in the same quantum state.

The operators s_+ and s_- commute with all the isospin operators. This can be verified by either calculating the commutators or by noting that a proton–neutron pair in the same quantum state has total isospin zero, and that the addition or removal of such a pair from a system cannot change the isospin. The operators s_+ and s_- do not commute with the baryon number since they clearly change the number of particles. A more convenient operator to use instead of the baryon number is

$$s_0 = \tfrac{1}{2}(B-1) = \tfrac{1}{2}(a_p^\dagger a_p - a_n a_n^\dagger) \,. \tag{5.2}$$

The three operators s_+, s_- and s_0 are now found to satisfy ordinary angular momentum commutation rules:

$$[s_0, s_+] = s_+ \,, \tag{5.3a}$$

$$[s_0, s_-] = -s_- \,, \tag{5.3b}$$

$$[s_+, s_-] = 2s_0 \,. \tag{5.3c}$$

We can thus call the s-operators quasispins. The commutation relation for the quasispin operators, the corresponding relation for the isospin operators and the fact that the quasispins and isospins commute with one another constitute the complete set of commutation relations for the Lie algebra:

$$[\tau_0, \tau_+[= \tau_+ \,, \tag{5.4a}$$

$$[\tau_0, \tau_-] = -\tau_- \,, \tag{5.4b}$$

$$[\tau_+, \tau_-] = 2\tau_0 \,, \tag{5.4c}$$

$$[\tau_i, s_k] = 0 \,. \tag{5.4d}$$

The Lie algebra of this set of six operators can therefore be expressed in terms of two sets of operators satisfying ordinary angular momentum commutation rules. There is therefore no difficulty in constructing all matrix elements of operators relevant to this algebra and in defining the structure of the multiplets. These are just direct products of two angular momentum multiplets.

There is no apparent physical meaning to this Lie algebra even after its extension to include space and spin by analogy with eq. (2.3). There is no system of nucleons where the addition or removal of a proton–neutron pair in the same quantum state has particular significance. The same quasispin algebra arises, however, in the treatment of pairing correlations in many-fermion systems. The relations (5.1), (5.2) and (5.3) can be applied to any two-fermion quantum states, not necessarily those of a proton and a neutron. Consider two quantum states denoted by k and $-k$ which might be two states with equal and opposite momentum $\pm\hbar k$. We can define quasispin operators

$$s_{+k} = a_k^\dagger a_{-k}^\dagger , \tag{5.5a}$$

$$s_{-k} = a_{-k} a_k , \tag{5.5b}$$

$$s_{0k} = \tfrac{1}{2}(a_k^\dagger a_k - a_{-k} a_{-k}^\dagger) . \tag{5.5c}$$

These quasispin operators now correspond physically to the addition or removal from the system of a pair of particles having equal and opposite momentum. Quasispin operators of the type (5.5) can be defined for any number of values of the momentum k and 'total quasispin' operators can be defined by summing the operators (5.5) over the set of states k under consideration:

$$S_+ = \sum_k s_{+k} , \tag{5.6a}$$

$$S_- = \sum_k s_{-k} , \tag{5.6b}$$

$$S_0 = \sum_k s_{0k} . \tag{5.6c}$$

These quasispin operators can be useful in considering pairing correlations because a simplified two-body pairing interaction can be expressed easily in terms of these operators. Consider, for example, the interaction

$$V = -G \sum_{kk'} a_k^\dagger a_{-k}^\dagger a_{-k'} a_{k'} \tag{5.7}$$

where the sum is over some particular set of states k. Such an interaction has been used in the BCS theory of superconductivity and in considering pairing correlations in complex nuclei. The interaction (5.7) is easily expressible in terms of the total quasispin operators (5.6)

$$V = -GS_+ S_- = -G\{S^2 - S_0^2 + S_0\} , \tag{5.8}$$

where the square of the total quasispin is of course just

$$S^2 = \tfrac{1}{2}(S_+ S_- + S_- S_+) + S_0^2 . \tag{5.9}$$

From the form (5.8) we can immediately characterize the eigenfunctions of the pairing interaction (5.7) and determine all its eigenvalues.

The eigenfunctions are just the simultaneous eigenfunctions of the operators S^2 and S_0 and the eigenvalues are obtained by direct substitution into eq. (5.8). The use of quasispin operators thus leads to a complete solution of the pairing problem in the so-called 'strong-coupling limit'; i.e. where the pairing force (5.7) is the dominating part of the Hamiltonian.

Note that in this particular case, the three quasispin operators associated with the addition or removal of pairs of particles to the system have a direct physical significance; however, the other three operators, which do not change the number of particles have no direct physical significance. The latter, analogous to isospin operators, simply move a particle from one quantum state to the other; i.e. from the state k to the state $-k$.

5.2. IDENTIFICATION OF THE LIE ALGEBRA

In considering the isospin Lie algebra we found that it was more natural to consider it as the algebra of the group of unitary transformations in two dimensions rather than that of three-dimensional rotations. This interpretation was easily generalized to include the case of three-dimensional unitary transformations. In order to consider generalizations of the Lie algebra (5.3) and (5.4) let us now specify more precisely the Lie algebra generated by the six bilinear products formed from all possible combinations of proton and neutron creation and annihilation operators. Experts on group theory will recognize immediately that a Lie algebra of six operators which can be separated into two independent angular momentum algebras must be the algebra of the four-dimensional rotation group. We can easily see this as follows. We first define the following linear combinations of the creation and annihilation operators

$$\gamma_1 = a_p^\dagger + a_p \,, \tag{5.10a}$$

$$\gamma_2 = i(a_p^\dagger - a_p) \,, \tag{5.10b}$$

$$\gamma_3 = a_n^\dagger + a_n \,, \tag{5.10c}$$

$$\gamma_4 = i(a_n^\dagger - a_n) \,. \tag{5.10d}$$

Examination of these operators shows that the square of any one of them is unity and that any pair of them anticommutes,

$$\gamma_i \gamma_j + \gamma_j \gamma_i = 2\delta_{ij} . \tag{5.11}$$

The operators (5.10) are four independent linear combinations of the proton and neutron creation and annihilation operators. The set of all independent, bilinear products of the operators (5.10) will be some linear combination of the set of all bilinear combinations of the proton and neutron operators; i.e. they will be some linear combination of the quasispin and isospin operators (5.3) and (5.4). Let us define the operators

$$L_{ij} = \tfrac{1}{2}\gamma_i \gamma_j \qquad (i \neq j) , \tag{5.12a}$$

$$L_{ij} = -L_{ji} , \tag{5.12b}$$

where eq. (5.12b) follows from the anticommutation relation (5.11). We see that there are indeed six independent operators of the type (5.12) and that they satisfy the commutation rules

$$[L_{ij}, L_{jk}] = L_{ik} = -L_{ki} \qquad (i \neq k) , \tag{5.13a}$$

$$[L_{ij}, L_{km}] = 0 \quad \text{if no two indices are equal.} \tag{5.13b}$$

The commutation rules (5.13) are just the natural extension of angular momentum commutation rules to four dimensions. We can consider the indices i, j, k and m as the directions of four axes in a four-dimensional space and the operator L_{ij} as the component of the angular momentum in the direction of the axis normal to the ij-plane. We note that for any set of three indices i, j and k we define a three-dimensional subspace of the four-dimensional space and the three operators L_{ij}, L_{jk} and L_{ki} satisfy the commutation rules of ordinary angular momentum operators in three dimensions.

We are now in a position to generalize these results to the case of an arbitrary number of creation and annihilation operators. Let us consider first the case of three states in the same way that we generalized isospin to obtain the Sakata model of elementary particles. We now wish to find the Lie algebra obtained by the most general set of independent, bilinear products of proton, neutron and Λ creation and annihilation operators. The preceding treatment is easily generalized by defining the following two new operators:

$$\gamma_5 = a_\Lambda^\dagger + a_\Lambda \,. \tag{5.14a}$$

$$\gamma_6 = \mathrm{i}(a_\Lambda^\dagger - a_\Lambda) \,, \tag{5.14b}$$

The definition of the operators L_{ij} (5.12) is easily extended to include the operators γ_5 and γ_6. By analogy with the previous case, we see that we have here the Lie algebra of the rotation group in six dimensions. The generalization of these results to an arbitrary number of quantum states is straightforward and evidently leads to the following result: Consider the set of fermion creation and annihilation operators $a_k^\dagger$ and a_k for n values of k. These may either be n different kinds of fermions or n different states of the same fermion. Then the set of all possible bilinear products of these creation and annihilation operators is a set of $n(2n-1)$ operators which constitute the Lie algebra for the rotation group in $2n$ dimensions. The set of all bilinear products of operators which do not change the number of particles, i.e. the product of a creation operator and an annihilation operator, is a set of n^2 operators which are a subset of the operators forming the Lie algebra of the rotation group in $2n$ dimensions. These operators constitute the Lie algebra of the group of unitary transformations in n dimensions. One linear combination of these n^2 operators is the operator of the total number of particles. It is therefore possible to separate the set of n^2 operators into the total number operator and a set of n^2-1 operators which constitute the Lie algebra of the special unitary or unimodular unitary group in n dimensions and is denoted by the letters SU_n.

5.3. SENIORITY

One might ask if it is possible to generalize the quasispin operators (5.1) which we obtain in the case $n=2$ and which allowed us to separate the Lie algebra of the four-dimensional rotation group into two independent angular momentum algebras, one of which, the isospin algebra, is just that of the unitary group which does not change the number of particles. One can see immediately that the exact analogy is not possible in general. In the case $n=2$ it was possible to find an operator which created a pair of particles in a state of isospin zero; i.e. belonging to the singlet multiplet of the

SU_2 group. In the Sakata model this is no longer possible. There is no state of two sakatons which belongs to the singlet multiplet of SU_3. There is therefore no hope of finding operators which create a pair of sakatons and which commute with all of the operators in the SU_3 algebra.

On the other hand, quasispin operators of the type (5.5) can be defined for any case where the number of quantum states n is even and can be grouped into pairs schematically denoted by k and $-k$.

An example of the use of these quasispin operators is in the seniority classification of nuclear states in the jj-coupling nuclear shell model. Let us consider a set of $2j+1$ creation operators $a^{\dagger}_{jm}$ creating a fermion in a state of total angular momentum j with projection m on the z-axis. The set of all possible bilinear products of creation and annihilation operators forms the Lie algebra of the rotation group in $2(2j+1)$ dimensions. Consider the multiplets corresponding to this Lie algebra which are relevant to the classification of states of any number of fermions distributed among these $2j+1$ single-fermion states. We find that all of the many-fermion states can be classified into two multiplets; one containing all the states having an odd number of particles and the other containing all the states having an even number of particles. This is evident since successive operation with the operators of the Lie algebra can connect any state having an even number of particles to any other state having an even number of particles, and similarly for the odd states. The unitary subgroup corresponding to those operators which do not change the number of particles is SU_{2j+1}. All possible states of a given number of particles in this j-shell correspond to a single multiplet of the group SU_{2j+1}.

Let us define quasispin operators analogous to (5.6)

$$S_+ = \tfrac{1}{2}\sum_m (-1)^{j-m} a^{\dagger}_{jm} a^{\dagger}_{j,-m}\,, \tag{5.15a}$$

$$S_- = \tfrac{1}{2}\sum_m (-1)^{j-m} a_{j,-m} a_{jm}\,, \tag{5.15b}$$

$$S_0 = \tfrac{1}{4}\sum_m a^{\dagger}_{jm} a_{jm} - a_{j,-m} a^{\dagger}_{j,-m}\,. \tag{5.15c}$$

where the additional factor of $\frac{1}{2}$ is inserted because the summation over all the values of m includes each term twice. The phase factor $(-1)^{j-m}$ is inserted by convention. We see that S_+ is an operator which creates a pair of particles in a state of total angular momentum $J=0$, while S_- annihilates such a pair of particles. Using these quasispin operators, we can define two quantum numbers by the eigenvalues of the total quasispin operator (5.9) and the component S_0.

The states of this system can be classified into quasispin multiplets. Let us examine the properties of these multiplets. We denote the states by $|S, S_0, \alpha\rangle$ where α represents all other quantum numbers independent of quasispin. From eq. (5.15c), the operator S_0 is simply related to the total number of particles, n.

$$S_0 = \tfrac{1}{2}(n-j-\tfrac{1}{2}), \tag{5.16a}$$

$$n = 2S_0+j+\tfrac{1}{2}. \tag{5.16b}$$

The eigenvalues of S_0 depend only on the number of particles and vary between $\pm\frac{1}{4}(2j+1)$, being zero in the middle of the shell. Since each eigenvalue of S_0 occurs only once in a given quasispin multiplet, these multiplets consist of states each having a different number of particles. For a multiplet of a given total quasispin S, S_0 varies from $-S$ to $+S$ and n varies between the limits

$$n_{\text{min}}(S_0=-S) = j+\tfrac{1}{2}-2S \equiv v, \tag{5.17a}$$

$$n_{\text{max}}(S_0=+S) = j+\tfrac{1}{2}+2S = 2j+1-v, \tag{5.17b}$$

where v is defined as the minimum number of particles occurring in a given quasispin multiplet and is called the seniority number. Since n and v are simply related to S_0 and S by eqs. (5.16b) and (5.17a), the states of the system can be labeled by the quantum numbers $|n, v, \alpha\rangle$ as well as $|S, S_0, \alpha\rangle$.

A state having the minimum number v of particles has $n=v$, or equivalently $S_0=-S$ and satisfies the relation

$$S_-|v, v, \alpha\rangle = S_-|S, -S, \alpha\rangle = 0. \tag{5.18}$$

The entire quasispin multiplet can then be built from this state by successive operation with the operator S_+; i.e. by successive addition of a pair of particles coupled to total angular momentum zero. The relation (5.18) implies that the lowest state in the multiplet contains no pairs coupled to angular momentum zero, since the pair annihilation operator S_- 'cannot find such a pair' in the state and gives zero.

In this quasispin or seniority classification system, the states are thus described as a state of v particles containing no pairs of zero angular momentum plus an arbitrary number $\frac{1}{2}(n-v)$ pairs of zero angular momentum. The states defined by this quasispin representation are automatically eigenfunctions of a pairing interaction analogous to (5.7)

$$V = -\tfrac{1}{4}G\sum_{mm'}(-1)^{2j-m-m'}a^\dagger_{jm}a^\dagger_{j,-m}a_{j,-m'}a_{jm'} \tag{5.19a}$$

$$= -GS_+S_- = -G(S^2-S_0^2+S_0)\,. \tag{5.19b}$$

Using eqs. (5.16) and (5.17), the eigenvalues of this pairing interaction are easily expressed in terms of the number of particles n and the seniority number v:

$$V = -\tfrac{1}{4}G(n-v)(2j+3-n-v)\,. \tag{5.20}$$

The seniority classification is particularly useful in cases where a pairing interaction plays a dominant role in the residual interaction, just as the SU_3 classification treated in § 4.4 is useful in cases where a quadrupole interaction is dominant.

Further interesting properties of the seniority classification are obtained from the commutation relations between the quasispin operators (5.15) and single creation and annihilation operators.

$$[S_+, a^\dagger_{jm}] = [S_-, a_{jm}] = 0\,, \tag{5.21a}$$

$$[S_+, a_{jm}] = (-1)^{j+m}a^\dagger_{j,-m}\,, \tag{5.21b}$$

$$[S_-, a^\dagger_{jm}] = (-1)^{j-m}a_{j,-m}\,, \tag{5.21c}$$

$$[S_0, a^\dagger_{jm}] = \tfrac{1}{2}a^\dagger_{jm}\,, \tag{5.21d}$$

$$[S_0, a_{jm}] = -\tfrac{1}{2}a_{jm}\,. \tag{5.21e}$$

These relations show that the pair of operators $(a^\dagger_{jm}; (-1)^{j-m} a_{j,-m})$ behave like 'two-component spinors in quasispin space'.

Any single creation or annihilation operator thus behaves like a component of a quasispin spinor. Products of two operators are linear combinations of quasispin scalars and quasispin vectors. Using the standard relations for combining two spins of one half to obtain the singlet and triplet states, we can construct the products

$$S(m, m') = \tfrac{1}{2}\sqrt{2}\{a^\dagger_{jm}(-1)^{j-m'} a_{j,-m'} - (-1)^{j-m} a_{j,-m} a^\dagger_{jm'}\}$$

$$= \tfrac{1}{2}\sqrt{2}\{(-1)^{j-m'} a^\dagger_{jm} a_{j,-m'} + (-1)^{j-m} a^\dagger_{jm'} a_{j,-m}\}\,, \quad (5.22a)$$

$$V_0(m, m') = \tfrac{1}{2}\sqrt{2}\{a^\dagger_{jm}(-1)^{j-m'} a_{j,-m'} + (-1)^{j-m} a_{j-m} a^\dagger_{jm'}\}\,, \quad (5.22b)$$

$$V_+(m, m') = a^\dagger_{jm} a^\dagger_{jm'}\,, \quad (5.22c)$$

$$V_-(m, m') = (-1)^{2j-m-m'} a_{j,-m} a_{j,-m'}\,, \quad (5.22d)$$

where $S(m, m')$ is a quasispin scalar and $V_0(m, m')$, $V_+(m, m')$ and $V_-(m, m')$ are the three components of a quasispin vector.

$$[S_+, S(m, m')] = [S_-, S(m, m')] = [S_0, S(m, m')] = 0\,, \quad (5.23a)$$

$$[S_+, V_+(m, m')] = [S_-, V_-(m, m')] = [S_0, V_0(m, m')] = 0\,, \quad (5.23b)$$

$$[S_0, V_\pm(m, m')] = \pm V_\pm(m, m')\,, \quad (5.23c)$$

$$[S_\pm, V_0(m, m')] = \sqrt{2}\, V_\pm(m, m')\,, \quad (5.23d)$$

$$[S_\pm, V_\mp(m, m')] = \sqrt{2}\, V_0(m, m')\,. \quad (5.23e)$$

For each set of values (m, m') a quasispin scalar $S(m, m')$ and a quasispin vector $V_k(m, m')$ can be constructed, except for the case $m = m'$ where the scalar exists but the vector vanishes. The scalar operators are of particular interest, since they commute with all the quasispin operators (5.15). Any function of these quasispin scalars is diagonal in the quasispin or seniority classification.

It is of interest to examine the behavior of the bilinear products of creation and annihilation operators also under rotations in ordinary three-dimensional configuration space; i.e. their commutation relations with the ordinary angular momentum operators. The set of creation operators $a^{\dagger}_{jm}$ transform among themselves like the components of an irreducible tensor of degree j, and similarly for the corresponding annihilation operators, with a suitable choice of phases. One can therefore construct irreducible tensors from bilinear products of operators by standard angular momentum couplings, for example

$$T_{kq} = (j\,j\,m\,m'|k\,q)\,a^{\dagger}_{m}a^{\dagger}_{m'}\,, \tag{5.24a}$$

$$T'_{kq} = (j\,j\,m\,m'|k\,q)(-1)^{j-m'}a^{\dagger}_{m}a_{-m'}\,, \tag{5.24b}$$

where T_{kq} and T'_{kq} are two irreducible tensors of degree k.

These irreducible tensors in ordinary configuration space can be expressed in terms of the quasispin vectors and scalars using the relations (5.22). The tensor T_{kq} (5.24a) contains only creation operators and therefore must be a component of a quasispin vector. A well-known property of the Clebsch-Gordan coefficients is that they do not change in absolute value with exchange of m and m', and do not change sign (for half-integral j) if k is odd, while they do change sign if k is even. Since $a^{\dagger}_{m}a^{\dagger}_{m'}+a^{\dagger}_{m'}a^{\dagger}_{m}=0$, T_{kq} vanishes for all odd values of k (another way of stating the well-known result that two identical fermions in the same j-shell can couple only to *even* values of the total angular momentum). By the same property of the Clebsch–Gordan coefficients, the tensors T'_{kq} (5.24b) consist of pairs of terms corresponding to given values of m and m' which are quasispin scalars $S(m, m')$ if k is odd and quasispin vectors $V_0(m, m')$ if k is even.

From these examples, we see that all odd tensors in configuration space constructed from bilinear products of creation and annihilation operators are quasispin scalars, while all even tensors are quasispin vectors. The three components of the quasispin vector are three even tensors, one like (5.24a) composed only of creation operators, one composed only of annihilation operators and one like (5.24b) composed of products of a creation and an annihilation operator. The quasispin scalar is always of the form (5.24b). Simple counting shows that the number of independent tensor components is equal to the total number of independent bilinear operator products and that this is in turn equal to the total number of independent quasispin vector and scalar components. Thus any bilinear product can be expanded in irreducible tensors. The odd tensors constructed from bilinear products are quasispin scalars, all such odd tensors commute with the quasispin operators and are diagonal in the seniority classification. Any interaction between particles which can be expressed as a function of only odd tensors is diagonal. The standard multipole expansion of a two-body interaction is an expansion in irreducible tensors which are just bilinear products (single-particle operators in the Schrödinger representation). Thus the contributions from the *odd* multipoles are diagonal in the seniority classification. This classification can be expected to be useful in treating an interaction which consists mainly of a pairing interaction (5.19) and an interaction constructed from odd tensors.

5.4. SYMPLECTIC GROUPS

The set of quasispin scalar operators $S(m, m')$ defined by eq. (5.22a) constitute a Lie algebra. They are the set of linear combinations of bilinear products of a creation and an annihilation operator which commute with all the quasispin operators. The commutator of any two such operators is thus some combination of the operators of the set. This algebra is included in the algebra of the group SU_{2j+1}, which includes all products of a creation and an annihilation operator. Let us attempt to identify this Lie algebra. The total number of independent operators is just the total number of pairs (m, m') which can be made from $2j+1$ values of m, including $m=m'$, but counting (m, m') and (m', m) only once. This therefore gives $\frac{1}{2}(2j+1)(2j+2)$, suggesting the rotation group in $2j+2$ dimensions. However, closer examination which is beyond the scope of this book indicates that this is only true for certain values of j. In general one obtains a new algebra which has the same number of operators as that of the rotation group in $2j+1$ dimensions but is different from it. This group is called the *symplectic group* in $2j+1$ dimensions.

Some basic properties of symplectic groups are obtained by noting that the set of operators (5.22a) commute with the quasispin operators (5.15) and can be considered to generate infinitesimal linear transformations on the set of creation or annihilation operators $a_{jm}^\dagger$ and a_{jm}, and that the set of creation operators $a_{jm}^\dagger$ can be considered as a vector in a space of $2j+1$ dimensions and the same for the corresponding annihilation operators. It is therefore natural to consider the symplectic group operators as performing transformations in a space of $2j+1$ dimensions, rather than of $2j+2$. The three quasispin operators (5.15) are bilinear products of components of vectors in this $2j+1$-dimensional space which are *invariant under the transformations of the symplectic group*; i.e. they commute with all the operators of the Lie algebra. However, it is evident by inspection that the quasispin operators are not ordinary scalar products of two vectors in this $2j+1$-dimensional space, as would be the case if the transformation which left them invariant were an ordinary rotation. The symplectic group is thus a kind of linear transformation in a vector space which leaves invariant a peculiar kind of bilinear product of two vectors.

Let us now specify more precisely the kind of vector product which is left invariant by symplectic transformations. From eqs. (5.15) we see that the components of each vector are classified into pairs, denoted by $+m$ and $-m$, and the product consists of terms in which the $+m$-component of one vector is multiplied by the $-m$ component of the second. This is also true for S_0, since a_{jm} *annihilates* angular momentum $+m$ and transforms like an object which *creates* angular momentum $(-m)$. The product also has the property of being *antisymmetric* with respect to interchange of the two vectors, or of the members of all the conjugate pairs $+m$ and $-m$. The general product of two vectors X and Y contains the combination $X_m Y_{-m} - X_{-m} Y_m$. We thus see that symplectic groups can be defined only in vector spaces having an *even* number of dimensions and that they leave invariant an antisymmetric product of two vectors requiring the classification of the dimensions of the space into conjugate pairs.

Let us now examine the behaviour of these symplectic transformations for the particular case of the *jj*-coupling shell model. We note immediately that for half-integral values of j, $2j+1$ is always even as is necessary for the definition of a symplectic group. Let us now consider the action of the operators (5.22a) on some many-particle state. These operators commute with all the quasispin operators (5.15) and cannot change the number of particles nor the seniority number v. Seniority defines a division of the number of particles into v particles in which there are no pairs coupled to angular momentum zero and $\frac{1}{2}(n-v)$ pairs of angular momentum zero. The symplectic group operators cannot change this division. They can change the states of the v particles which are not coupled to angular momentum zero, but they cannot change the number of pairs coupled to zero. This is just what is implied by their commutation with the quasispin operators: *they leave invariant a two-particle state with angular momentum zero.*

The multiplets generated by these operators of the symplectic group in $2j+1$ dimensions thus consist of states all having the same number of particles and the same seniority, but having the v unpaired particles in different states. They are complementary in a sense to the quasispin multiplets, which also consist of states having the same seniority, but differ in having different total numbers of particles (i.e. different numbers of added zero pairs), while the v unpaired particles are left essentially in the same state (except for Pauli principle effects which restrict overlap of the wave function of the v particles with those particles in the added pairs).

We shall now see that the quasispin operators themselves, (5.15), are more naturally interpreted as the Lie algebra of the symplectic group in two dimensions, rather than that of the three-dimensional rotation group. As in the case of isospin, there is no physical three-dimensional space associated with these quasispin operators in any simple way. Rather, the quasispin operators generate infinitesimal transformations mixing the components of two-component objects; namely the quasispin spinors $(a^{\dagger}_{jm}; (-1)^{j-m}a_{j,-m})$. This is also analogous to isospin, which generates infinitesimal transformations between proton and neutron states which can be considered as spinors in a three-dimensional isospin space. However, it is more natural in both cases to consider these two-component objects as vectors in a two-dimensional space, rather than spinors in a three-dimensional space which has no direct physical interpretation. In the case of the quasispin operators (5.15) it is evident that the transformations generated in the two-dimensional space are symplectic, rather than unitary as in isospin. The product of two vectors in the two-dimensional space which remains invariant under the quasispin transformations is just the product $S(m, m')$ defined by eq. (5.22a) which commutes with all quasispin operators. This product is not the ordinary scalar product of two vectors, which remains invariant under unitary transformations, but is just the peculiar antisymmetric product which remains invariant under symplectic transformations. Since there are only two dimensions in the space, the two dimensions are conjugate to one another, and the antisymmetric product of two vectors is formed by taking the product of one component of the first vector with the *other* component of the second and taking the difference between the two terms formed in this way.

The symplectic group in n dimensions is usually denoted as Sp_n. We see that the Lie algebra of Sp_2 is the same as that of SU_2 and the ordinary angular momentum algebra. This is another example of a Lie algebra of rank one which is the same as that of angular momentum.

5.5. SENIORITY WITH NEUTRONS AND PROTONS. THE GROUP Sp_4

Let us now consider the extension of the quasispin operators (5.15) to the case where there are both neutrons and protons in the same j-shell. Let $a^\dagger_{pm}$ and $a^\dagger_{nm}$ denote creation operators for a proton and a neutron respectively in a state of total angular momentum j with projection m on the z-axis. The index j is omitted for convenience, since all operators used in any specific case refer to the same value of j. There are now $2(2j+1)$ single fermion states, and $4(2j+1)$

different single fermion creation and annihilation operators. The set of all possible bilinear products thus forms the Lie algebra of the rotation group in $4(2j+1)$ dimensions, and the subset of bilinear products which does not change the number of particles forms the Lie algebra of the unitary group in $2(2j+1)$ dimensions. We note that the isospin operators (2.3) are a particular subset of the operators in the unitary group in $2(2j+1)$ dimensions.

There are three ways of making a pair of particles with zero angular momentum: a proton–proton pair, a proton–neutron pair, or a neutron–neutron pair. In the isospin formalism a pair of nucleons with zero angular momentum is a member of an isospin triplet, having $T=1$ and three possible states with $T_0=+1$, 0 and -1. The quasispin operators (5.15) can be extended to this case by defining three quasispins, one for proton–proton pairs, one for neutron–neutron pairs and one for neutron–proton pairs. These three quasispins do not commute among one another. The commutator of an operator creating a proton pair with one annihilating a neutron–proton pair is an operator creating one proton and annihilating one neutron and turns out to be just the isospin operator τ_+. Including all commutators leads to a Lie algebra of ten operators, three pair creation operators, three pair annihilation operators, three isospin operators and the total number operator measured from the middle of the shell, like S_0, eq. (5.15c). Since $10=5\times 4/2$, the algebra of the five-dimensional rotation group is suggested and also that of the four-dimensional symplectic group. It turns out that the two algebras are identical, but that the symplectic group has a simpler interpretation, as in the case of the quasispins (5.15).

Let us now investigate the algebra in more detail. We denote creation operators for a pair of nucleons in the state $T=1$, $T_0=+1$, 0 and -1 respectively by $A_+^\dagger$, $A_0^\dagger$ and $A_-^\dagger$, the corresponding annihilation operators by A_+, A_0 and A_-, and half the number of particles measured from zero in the middle of the shell by N_0:

$$A_+^\dagger = \tfrac{1}{2}\sum_m (-1)^{j-m} a_{pm}^\dagger a_{p,-m}^\dagger \tag{5.25a}$$

$$A_0^\dagger = \tfrac{1}{2}\sqrt{2}\sum_m (-1)^{j-m} a_{pm}^\dagger a_{n,-m}^\dagger \tag{5.25b}$$

$$A_-^\dagger = \tfrac{1}{2}\sum_m (-1)^{j-m} a_{nm}^\dagger a_{n,-m}^\dagger \tag{5.25c}$$

$$A_+ = \tfrac{1}{2}\sum_m (-1)^{j-m} a_{p,-m} a_{pm} \tag{5.25d}$$

$$A_0 = \tfrac{1}{2}\sqrt{2}\sum_m (-1)^{j-m} a_{n,-m} a_{pm} \tag{5.25e}$$

$$A_- = \tfrac{1}{2}\sum_m (-1)^{j-m} a_{n,-m} a_{nm} \tag{5.25f}$$

$$\tau_+ = \sum_m a_{pm}^\dagger a_{nm} \tag{5.25g}$$

$$\tau_- = \sum_m a_{nm}^\dagger a_{pm} \tag{5.25h}$$

$$\tau_0 = \tfrac{1}{2}\sum_m a_{pm}^\dagger a_{pm} - a_{nm}^\dagger a_{nm} \tag{5.25i}$$

$$N_0 = \tfrac{1}{2}\sum_m a_{pm}^\dagger a_{pm} - a_{nm} a_{nm}^\dagger$$

$$= (\tfrac{1}{2}\sum_m a_{pm}^\dagger a_{pm} + a_{nm}^\dagger a_{nm}) - \tfrac{1}{2}(2j+1) . \tag{5.25j}$$

The additional normalization factor $\sqrt{2}$ is included in $A_0^\dagger$ and A_0 because protons and neutrons are not identical particles. With this normalization the three pair creation operators constitute an isospin triplet. These operators can be shown to satisfy the commutation relations

$$[\tau_\pm, A_\mp^\dagger] = +\sqrt{2}A_0^\dagger; \quad [\tau_0, A_\pm^\dagger] = \pm A_\pm^\dagger; \quad [\tau_\pm, A_0^\dagger] = +\sqrt{2}A_\pm^\dagger$$

$$[\tau_\pm, A_\pm] = -\sqrt{2}A_0; \quad [\tau_0, A_\pm] = \mp A_\pm; \quad [\tau_\pm, A_0] = -\sqrt{2}A_\mp$$

$$[\tau_\pm, A_\pm^\dagger] = [\tau_\pm, A_\mp] = [\tau_0, A_0^\dagger] = [\tau_0, A_0] = [\tau_0, N_0] = = [\tau_\pm, N_0] = 0$$

$$[N_0, A_\pm^\dagger] = +A_\pm^\dagger; \quad [N_0, A_0^\dagger] = +A_0^\dagger$$

$$[N_0, A_\pm] = -A_\pm; \quad [N_0, A_0] = -A_0$$

$$[A_\pm^\dagger, A_\pm] = N_0 \pm \tau_0; \quad [A_0^\dagger, A_0] = N_0; \quad [A_\pm^\dagger, A_0] = [A_0^\dagger, A_\mp] = \tfrac{1}{2}\sqrt{2}\,\tau_\pm$$

$$[A_+^\dagger, A_-^\dagger] = [A_\pm^\dagger, A_0^\dagger] = [A_+, A_-] = [A_\pm, A_0] = [A_\pm^\dagger, A_\mp] = 0$$

$$[\tau_+, \tau_-] = 2\tau_0; \qquad [\tau_0, \tau_\pm] = \pm\tau_\pm .$$

The algebra is evidently of rank two. We can choose N_0 and τ_0 to be diagonal. The remaining eight operators are then represented diagrammatically as shown in Fig. 5.1.

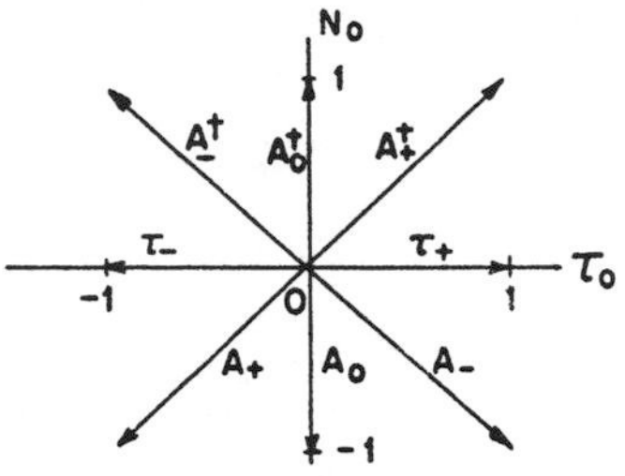

Fig. 5.1

The diagram is easily constructed by noting that the pair creation and annihilation operators constitute isospin triplets which change N_0 by ± 1 and that the isospin operators do not change N_0.

From this set of operators (5.25) four independent sets of quasispin operators can be found which satisfy angular momentum commutation rules; namely the proton pair, neutron pair, and proton–neutron pair quasispins, and isospin. These correspond to directions on the diagram of Fig. 5.1 of $\pm 45°$, vertical and horizontal. The proton pair and neutron pair operators commute with one another, but none of the other pairs of quasispins commute. The corresponding multiplet diagrams for this group form two-dimensional lattices whose lattice vectors are those of Fig. 5.1. Like the multiplets of the group SU_3 there may be several states occurring at a given point on a diagram, and additional quantum numbers must be found to distinguish between them. Since the proton pair and neutron pair quasispins commute, two additional quantum numbers are available, the total neutron quasispin and the total proton quasispin. These can be shown to be sufficient for classifying the states within the multiplet. Unfortunately, this classification is not convenient for nuclear applications where the interactions conserve isospin to a good approximation. The isospin does not commute with the proton pair and neutron pair quasispins and is not diagonal in this classification. If the total isospin T^2 is chosen to be diagonal, there is no simple way to find another quantum number to specify the states. The detailed multiplet structure can be determined by methods analogous to those described for the group SU_3 in Chapter 3 and Appendix A.

If the isospin operators are considered to be the Lie algebra for the three-dimensional rotation group, the complete set of ten operators (5.25) can be shown to constitute the Lie algebra for a rotation group in a five-dimensional space including, as a subspace, the three-dimensional space of the isospin rotations. Operators L_{ij} satisfying the commutation relations (5.13) for a rotation group can be defined as follows:

$$\tau_0 = L_{12}\,, \tag{5.26a}$$

$$\tau_\pm = L_{23} \pm \mathrm{i}L_{31}\,, \tag{5.26b}$$

$$A_\pm^\dagger = L_{41} \pm \mathrm{i}L_{42} + \mathrm{i}(L_{51} \pm \mathrm{i}L_{52})\,, \tag{5.26c}$$

$$A_\pm = L_{41} \mp \mathrm{i}L_{42} - \mathrm{i}(L_{51} \mp \mathrm{i}L_{52})\,, \tag{5.26d}$$

$$A_0^\dagger = L_{43} + \mathrm{i}L_{53}\,, \tag{5.26e}$$

$$A_0 = L_{43} - \mathrm{i}L_{53}\,, \tag{5.26f}$$

$$N_0 = L_{54}\,. \tag{5.26g}$$

With respect to isospin transformations, the ten operators can be classified as one isoscalar and three isovectors, with one isovector being the isospin operators themselves.

Isospin is more naturally considered as the Lie algebra of the group SU_2 rather than that of three-dimensional rotations, and there is no simple interpretation for any five-dimensional space in which the operators (5.25) generate transformations. These operators are simply interpreted as generating symplectic transformations in a four-dimensional space by combining the two two-dimensional symplectic transformations of the neutron pair and proton pair quasispins. Let us define quasispin scalars $S_n(m, m')$ and $S_p(m, m')$

for the neutron and proton pairs respectively, analogous to (5.22a)

$$S_{\rm n}(m, m') = a^\dagger_{{\rm n}m}(-1)^{j-m'} a_{{\rm n},-m'} - (-1)^{j-m} a_{{\rm n},-m} a^\dagger_{{\rm n},m'} , \quad (5.27a)$$

$$S_{\rm p}(m, m') = a^\dagger_{{\rm p}m}(-1)^{j-m'} a_{{\rm p},-m'} - (-1)^{j-m} a_{{\rm p},-m} a^\dagger_{{\rm p},m'} . \quad (5.27b)$$

The sum of these two quasispin scalars $S_{\rm n}(m, m') + S_{\rm p}(m, m')$ is seen to be an isospin scalar,

$$[\tau_\pm, (S_{\rm n} + S_{\rm p})] = [\tau_0, (S_{\rm n} + S_{\rm p})] = 0 . \quad (5.28)$$

The operator $S_{\rm n} + S_{\rm p}$ commutes with the neutron pair and proton pair quasispins as well as with isospin. It therefore commutes with all the ten operators (5.25), since the neutron–proton pair operators can be made from commutators of isospin and the other quasispins. The operator $S_{\rm n} + S_{\rm p}$ thus remains invariant under the transformations generated by the operators (5.25). The operator $S_{\rm n} + S_{\rm p}$ has the exact form of the antisymmetric product of two vectors which remains invariant under symplectic transformations, if we define $(a^\dagger_{{\rm n}m}; a^\dagger_{{\rm p}m}; (-1)^{j-m} a_{{\rm n},-m}; (-1)^{j-m} a_{{\rm p},-m})$ as the components of a vector in a four-dimensional space.

Since this Lie algebra is of rank 2, there are two independent Casimir operators which commute with all the operators and which can be used to label the multiplets. As in the case of SU_3, the eigenvalues of these operators are probably complicated. However, two quantum numbers which should be related to these Casimir operators are easily found. By analogy with the simple seniority case of § 5.3, we can expect the states within a given multiplet to be generated from a state containing a certain number v of particles in which no pair is coupled to angular momentum zero; i.e. all the pair annihilation operators give zero when acting on this state. By successive operation with the operators (5.25) the states of the multiplet are generated, all containing v particles with no pairs coupled to zero and a number of additional pairs. The initial state of v particles can be chosen to have a definite total isospin t_0. The two quantum numbers v and t_0 thus characterise the multiplet. The quantum number v is again called the seniority and t_0 is called the reduced isospin.

The result for the simple seniority case that any interaction constructed from odd tensors in ordinary configuration space is diagonal in the seniority classification is easily generalized to the case of seniority with neutrons and protons. The quantities $S_p(m, m')$ and $S_n(m, m')$ contain only odd tensors, as shown by eq. (5.24). Any odd tensor which is independent of the charge (i.e. an isospin scalar) is expressible in terms of these quantities and therefore commutes with all the operators (5.26) of Sp_4. An interaction constructed from these odd tensors is thus diagonal in the classification of seniority and reduced isospin.

5.6. LIE ALGEBRAS OF BOSON OPERATORS. NON-COMPACT GROUPS

Let us now consider the Lie algebra formed by bilinear products of boson creation and annihilation operators which can change the number of particles. The simplest case is that of a single boson state, with creation operator $a^\dagger$ and annihilation operator a. This corresponds to the case of a single harmonic oscillator. There are three independent bilinear products. For our purposes it is convenient to define them as follows:

$$t_1 = \tfrac{1}{4}(a^\dagger a^\dagger + aa)\,, \tag{5.29a}$$

$$t_2 = -\tfrac{1}{4}\mathrm{i}(a^\dagger a^\dagger - aa)\,, \tag{5.29b}$$

$$t_3 = \tfrac{1}{4}(a^\dagger a + aa^\dagger)\,. \tag{5.29c}$$

Since there are only three operators, we expect to find a Lie algebra of rank one which is the same as the angular momentum algebra. However, the commutators of the three operators (5.29) differ from those of angular momenta by a sign which cannot be eliminated by any redefinition of phases.

$$[t_1, t_2] = -\mathrm{i}t_3\,, \tag{5.30a}$$

$$[t_2, t_3] = \mathrm{i}t_1\,, \tag{5.30b}$$

$$[t_3, t_1] = \mathrm{i}t_2\,. \tag{5.30c}$$

It is possible to obtain operators satisfying commutation rules formally like angular momenta by defining

$$j_1 = \mathrm{i}t_1 \,, \tag{5.31a}$$

$$j_2 = \mathrm{i}t_2 \,, \tag{5.31b}$$

$$j_3 = t_3 \,. \tag{5.31c}$$

Then

$$[j_1, j_2] = \mathrm{i}j_3 \;; \qquad [j_2, j_3] = \mathrm{i}j_1 \;; \qquad [j_3, j_1] = \mathrm{i}j_2 \,. \tag{5.32}$$

However, the operators j_1 and j_2 are *antihermitean* rather than being hermitean as in the case of angular momentum operators. We can therefore use only those results from angular momentum which depend on the commutation rules, and not those which use the hermiticity of the operators. This leads to interesting differences.

We can define a 'total angular momentum'

$$j^2 = j_1^2 + j_2^2 + j_3^2 = t_3^2 - t_1^2 - t_2^2 \,. \tag{5.33}$$

As in angular momentum, j^2 commutes with all the operators (5.31), and we can choose j^2 and j_3 to be simultaneously diagonal. However, t_1 and t_2 are hermitean, rather than j_1 and j_2. Eq. (5.33) shows that the eigenvalue of j^2 is *less* than the eigenvalue of j_3^2 for any state which is a simultaneous eigenfunction of j^2 and j_3.

We can again define step operators $j_1 \pm \mathrm{i}j_2$. However, the hermitean conjugate of $j_1 + \mathrm{i}j_2$ is $-(j_1 - \mathrm{i}j_2)$! This leads to a modified step operator equation, since hermiticity is used in determining the coefficients. Let $|j^2, m\rangle$ denote a simultaneous eigenfunction of j^2 and j_3 with eigenvalues j^2 and m. Then by a procedure analogous to that for angular momentum,

$$(j_1 \pm \mathrm{i}j_2)|j^2, m\rangle = \sqrt{m(m \pm 1) - j^2}\, |j^2, m \pm 1\rangle \,. \tag{5.34}$$

The reversal of the sign under the square root in (5.34) is consistent with the observation that $m^2 \geqslant j^2$, rather than vice versa. In eq. (5.34), j^2 means the eigenvalue of the operator and corresponds to the familiar $j(j+1)$ for angular momenta. However, we have not yet determined the eigenvalues of j^2 for this case.

The eigenvalues of j^2 can be found by examining the end of the multiplet. Since $m^2 \geqslant j^2$, there must be *a minimum* value of m^2 for a given value of j^2. As in angular momentum the coefficient under the square root must vanish at the end of the multiplet to prevent creation of states beyond the end. Here, however, this condition leads to the relation

$$j^2 = |m|_{\min}\{|m|_{\min} - 1\} . \tag{5.35}$$

It is thus convenient to define a number j, such that the eigenvalue of j^2 is $j(j-1)$. We then have allowed values of m going in steps of unity from $+j$ to $+\infty$ and from $-j$ to $-\infty$. The Lie algebra does not introduce any restrictions on eigenvalues of j, which could in principle have a continuous spectrum. However, the explicit form of t_3, eq. (5.29), introduces restrictions, because its eigenvalues are known. They are all positive, thus eliminating the negative eigenvalues, and they are *quarter-integral*; i.e. $\frac{1}{4}$, $\frac{3}{4}$, $\frac{5}{4}$, etc., because of the factor $\frac{1}{4}$ in the definition of t_3.

Evaluation of the operator j^2 explicitly in terms of the creation and annihilation operators reveals that it is a very trivial operator; a c-number equal to $-\frac{3}{16}$. This corresponds to two values of j, as the relation $j(j-1) = -\frac{3}{16}$ has two solutions, $j = \frac{1}{4}$ and $j = \frac{3}{4}$. The states with $j = \frac{1}{4}$ have $m = \frac{1}{4}, \frac{5}{4}, \frac{9}{4}, \ldots$ and those with $j = \frac{3}{4}$ have $m = \frac{3}{4}, \frac{7}{4}, \frac{11}{4}, \ldots$. These are just the states of a harmonic oscillator with even and odd parity respectively, or even and odd numbers of quanta.

The algebra becomes less trivial if a set of n harmonic oscillators is considered. This might be several particles moving in an oscillator potential or a multidimensional harmonic oscillator. If we let $a_k^\dagger$ and a_k be creation and annihilation operators for the kth oscillator, we can again obtain a Lie algebra like (5.30) by defining corresponding operators summed over k:

$$T_1 = \sum_{k=1}^{n} \tfrac{1}{4}(a_k^\dagger a_k^\dagger + a_k a_k) , \tag{5.36a}$$

$$T_2 = \sum_{k=1}^{n} -\tfrac{1}{4}\mathrm{i}(a_k^\dagger a_k^\dagger - a_k a_k) , \tag{5.36b}$$

$$T_3 = \sum_{k=1}^{n} \tfrac{1}{4}(a_k^\dagger a_k + a_k a_k^\dagger) , \tag{5.36c}$$

$$[T_1, T_2] = -\mathrm{i}T_3; \quad [T_2, T_3] = +\mathrm{i}T_1; \quad [T_3, T_1] = +\mathrm{i}T_2 , \tag{5.36d}$$

$$J^2 = T_3^2 - T_1^2 - T_2^2 . \tag{5.36e}$$

The operator J^2 is no longer a c-number for this case. The eigenvalues of J^2 have the form $J(J-1)$ where J is an integer or half-integer if n is even and a quarter-integer if n is odd. For each value of J which occurs, a multiplet is defined which contains an infinite number of states with eigenvalues of T_3 starting at $T_3 = J$ and increasing without limit in unit steps.

For the case where the set of oscillators represent a system of particles moving in an oscillator potential, these multiplets have a simple physical interpretation as collective vibrational bands. Consider a system of particles moving classically in a harmonic oscillator potential. Let them all be placed initially at the origin and given arbitrary initial velocities at the same time. The particles execute independent oscillations with random amplitude and direction, but all with the same frequency and in phase. An observer might describe the motion as an expanding and contracting cloud of particles moving in a collective oscillation at double the oscillator frequency. In quantum mechanics, one cannot introduce correlations by choosing initial conditions in space and time. However, a state having collective properties can be constructed by choosing an appropriate linear combination of degenerate states. A system of particles moving in a harmonic oscillator potential has a highly degenerate spectrum and different types of correlations are described by choosing different linear combinations. The choice of those states which are eigenfunctions of the operator J^2 (5.33) can be shown to exhibit the collective dilatational oscillation or 'breathing mode' analogous to the above classical example. The lowest state in a band having a given value of J^2 describes some motion of the particles with no collective dilatation present. The higher states in the multiplet form a vibrational band based on this 'intrinsic' state, with a 'phonon' energy of just double the oscillator level spacing.

Let us now examine the peculiar group of transformations defined by this algebra of angular momentum operators with one wrong sign. Such a sign difference is associated with rotations in a space which is not Euclidean, but is of a Minkowsky type. The transformations generated by the operators (5.33) are like Lorentz transformations in a three-dimensional space having a metric $x_3^2 - x_1^2 - x_2^2$.

However, there is again no three-dimensional space associated with this problem in any simple way. If we consider the pair of operators $(a_k^\dagger, a_k)$ as the components of a vector in a two-dimensional space, we find that the antisymmetric product of two such vectors $a_k^\dagger a_{k'} - a_k a_{k'}^\dagger$ commutes with the operators (5.33) and remains invariant under the transformation. The transformations are thus more naturally considered to be symplectic transformations rather than rotations or Lorentz transformations. We thus again encounter the group Sp_2. However, it is different from the group Sp_2 discussed in § 5.4 for seniority because of the difference in sign in the commutators.

We note another peculiar feature resulting from the 'wrong sign' in the commutation rules. The multiplets all contain an infinite number of states. Groups having this property are called 'non-compact' groups, in contrast to those we have considered up to this point, which have multiplets of finite size and are called compact. If we are constructing a Lie algebra with bilinear products of second quantized operators, we shall obtain algebras of non-compact groups whenever we include boson operators which change the number of particles. The multiplets are all infinite, because there is nothing to prevent operating again and again on a state with an operator which adds a pair of bosons to the system. With fermion operators, the procedure of adding pairs of particles must eventually end if there are only a finite number of states available. With operators which do not change the number of particles, there are again only a finite number of states for the many-particle system and the multiplets must be finite, whether the particles are bosons or fermions.

This treatment is easily generalized to consider the case of n-boson states and all the bilinear products which can be constructed from the creation and annihilation operators. The total number of bilinear products is $n(2n+1)$, rather than $n(2n-1)$ for the corresponding fermion case. The additional $2n$ operators are just the squares $(a_k^\dagger)^2$ and a_k^2 which vanish for fermions but are perfectly good operators for bosons. The number of operators $n(2n+1)$ suggests the rotation group in $2n+1$ dimensions or the symplectic group in $2n$ dimensions. From the simple case of $n=1$, we expect that Sp_{2n} is the correct assignment, and this can be shown to be the case.

Note the similarity in character between the operators (5.33) and the pairing quasispins. The three operators include a pair creation operator, a pair annihilation operator and a number operator. This analogy persists in the cases of higher dimensions, as is to be expected since it also involves the algebra of the group Sp_{2n}. For the case $n=2$, we can consider a system of particles moving in a two-dimensional harmonic oscillator potential. There is a direct parallel to the neutron–proton seniority problem. Instead of creating neutrons and protons, one creates oscillator quanta in the x- and y-directions. There are three pair creation operators xx, yy and xy, analogous to pp, nn and pn. The SU_2 group formed by the operator products which do not change the number of particles, discussed in Chapter 4, is analogous to the isospin group, and there is again the operator of the number of particles. However, the group defined by the boson operators is again non-compact, as indicated by the possibility of adding pairs ad infinitum. If the algebra of Sp_4 for bosons is examined, differences in sign from the fermion case are again encountered. Since the algebra of Sp_4 is the same as that of the five-dimensional rotation group, one can examine the commutators for the boson case and find that it corresponds to the Lie algebra of the De Sitter group, a group like the Lorentz group in a five-dimensional space having a metric $x_1^2+x_2^2-x_3^2-x_4^2-x_5^2$.

5.7. THE GENERAL CLASSIFICATION OF LIE ALGEBRAS OF BILINEAR PRODUCTS

We have now identified the Lie algebras generated by taking all possible bilinear products of second-quantized operators for a finite number n of states. Both fermions and bosons have been considered, and algebras consisting only of products which do not change the number of particles as well as those including products which change the number of particles. The four possible cases are summarized in Table 5.1.

TABLE 5.1

Statistics	*Number of particles*	*Number of bilinear products*	*Lie algebra*	*Compact*
fermions	unchanged	n^2	U_n	yes
fermions	changed	$n(2n-1)$	R_{2n}	yes
bosons	unchanged	n^2	U_n	yes
bosons	changed	$n(2n+1)$	Sp_{2n}	no

CHAPTER 6

PERMUTATIONS, BOOKKEEPING AND YOUNG DIAGRAMS

Permutation symmetry is useful in treating states of many-particle systems. It is also useful in combining multiplets of a particular Lie algebra and in determining the structure of multiplets which can be built up from smaller multiplets. The basic principle behind this usage is that the operators of a Lie algebra defined for a system of identical particles are symmetric with respect to permutation of the particles. Operators which produce permutations of particles thus commute with the operators of the Lie algebra. States of systems of identical particles belonging to the same multiplet of some Lie algebra must all have the same permutation symmetry. The use of this principle is illustrated in Appendix A in constructing the SU_3 multiplets.

A familiarity with permutation groups is useful in treating such problems, and a convenient tool for handling permutation problems is the Young diagram or tableau. The purpose of this chapter is to give the reader a general impression of permutation techniques and the meaning and use of Young diagrams without going into the detailed theory of the permutation group. The isospin couplings in a many-nucleon system furnishes a convenient example.

The use of the isospin formalism for systems consisting only of nucleons can in some respects be considered as merely a matter of bookkeeping. The physical principle behind isospin is the charge independence of nuclear forces. This states that as far as strong interactions are concerned all nucleons are equivalent, whether they be neutrons or protons. However, life is not quite that simple. Although all nucleons are equivalent, some nucleons are more equivalent than others. The Hamiltonian may not care whether a nucleon is a neutron or a proton, but the Pauli exclusion principle does care. A wave function describing a system of nucleons must be antisymmetric with respect to exchanges of neutrons or with respect to exchanges of protons, but the Pauli principle does not care about exchanging a neutron and a proton. Given a wave function which describes a particular state of motion of a system of neutrons, the Hamiltonian says that we can change neutrons to protons and vice versa at will without changing the forces and therefore without changing the motion of the particles. However, if we start with a possible state of motion and change neutrons into protons and vice versa we may end up with a state which violates the Pauli principle.

Some system of bookkeeping is necessary in order to keep track of the Pauli principle. This bookkeeping is simple only in the two-nucleon system, which is treated in all elementary courses. The bookkeeping of the three-nucleon system is already beyond the scope of most simple treatments. The isospin formalism offers a method for avoiding complicated bookkeeping.

Let us first consider the bookkeeping of a two-nucleon system. The space-spin states can be either symmetric or antisymmetric. The Pauli principle requires that if the two nucleons are really equivalent, i.e. they are either both neutrons or both protons, then only the antisymmetric states are allowed. However, if the two-nucleon system consists of a neutron and a proton, then either symmetric or antisymmetric states are allowed.

These same requirements are stated as follows using the isospin formalism: The nucleon is represented as an isospin doublet with $T=\frac{1}{2}$ and one specifies the state of the nucleon by giving not only its space and spin quantum numbers but the eigenvalue of τ_0 which tells whether it is a neutron or a proton. One then requires ('generalized Pauli principle') the total wave function to be antisymmetric in space-spin and isospin. It is convenient to use wave functions which are products of a space-spin part and an isospin part. The overall wave function must be antisymmetric. Thus, if the space part is antisymmetric the isospin part must be symmetric; i.e. it must have $T=1$. There are three states with $T=1$: a neutron–neutron state, a proton–proton state, and the symmetric proton–neutron state. On the other hand, if the space-spin part is symmetric then the isospin part must be antisymmetric; i.e. it must have $T=0$. There is only a single $T=0$ state, which must have one neutron and one proton. We see again that the neutron–neutron and proton–proton states must be antisymmetric in space and spin, whereas the neutron–proton states can be either symmetric or antisymmetric.

The two-nucleon system is simple because we can always talk about states as being either symmetric or antisymmetric. In a three-nucleon system other symmetries arise. It is easy to write down a three-particle wave function which is either completely symmetric or completely antisymmetric. However, there are also three-particle states which are partly symmetric and partly antisymmetric; i.e. they may be symmetric with respect to the interchange of one pair of particles, but antisymmetric with respect to the interchange of another. The generalized Pauli principle in the isospin formalism requires that if the spacial part of the wave function has one of these mixed symmetries, the isospin part should have some kind of complementary mixed symmetry so that the overall wave function is antisymmetric. The complicated bookkeeping problems which arise here are avoided in the isospin formalism by use of coupling rules which simply add isospins like angular momenta.

A convenient bookkeeping shortcut to keep track of symmetries is the use of Young diagrams. The following pedestrian description of how these diagrams can be used is not intended to be rigorous, but gives an idea of how the thing works. Let us represent each nucleon in the system by a square □. We represent a two-nucleon system by two squares. We write the two squares as a vertical array $\begin{array}{|c|}\hline \ \\ \hline \ \\ \hline\end{array}$ to represent a two-nucleon system with an antisymmetric isospin function, i.e. with $T=0$. We use a horizontal array of two squares to represent two nucleons □□ with a symmetric isospin function, i.e. in the state with $T=1$. A state of three nucleons which is symmetric in isospin, i.e. has isospin $\frac{3}{2}$, is represented by three squares in a horizontal array.

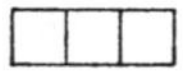

A state of n nucleons which is completely symmetric in isospin and therefore has $T=\frac{1}{2}n$ is represented by a horizontal array of n squares.

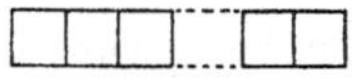

In any system of three or more nucleons there must be at least one isospin state occupied by two nucleons, since there are only two possible states for the isospin quantum number of a single nucleon. Thus, we cannot have a system of three or more nucleons which has a wave function that is completely antisymmetric in isospin. Although we can have as many squares as we please lined up in a horizontal direction in a diagram indicating symmetric states in isospin, we can never have more than two squares in a vertical line indicating a totally antisymmetric state.

We now have diagrams representing all possible states which are either totally antisymmetric or totally symmetric in isospin; i.e. the two-nucleon state with $T=0$ or the n-nucleon state with $T=\frac{1}{2}n$. We know that in general for an n-nucleon state we can have values of T of $\frac{1}{2}n$, $\frac{1}{2}n-1$, $\frac{1}{2}n-2$, etc. For example, the three-nucleon system can have values of T of either $\frac{3}{2}$ or $\frac{1}{2}$. We shall represent the state of three nucleons with $T=\frac{1}{2}$ as follows:

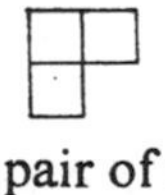

This indicates that there are a pair of nucleons in the antisymmetric state, $T=0$, and an additional nucleon. A state of six nucleons having isospin 1 would be represented by

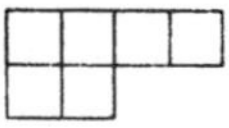

Here we have two pairs of nucleons which are coupled to $T=0$ and represented by vertical arrays of two squares, plus two additional nucleons coupled to $T=1$, and represented by the horizontal array. In general, we can represent an n-nucleon state having isospin T as follows: First we have $\frac{1}{2}n-T$ pairs of particles coupled to $T=0$ and represented by vertical arrays of two squares placed next to one another. The remaining $2T$ nucleons are coupled to total isospin T and represented by a horizontal array of T squares which are added on to the diagram at the upper right. Each state is thus characterized by two numbers: the number of nucleons n and the total isospin T.

Instead of the total number of particles n we can also use the total number of antisymmetric pairs $\frac{1}{2}n-T$. It is conventional to define the quantum numbers λ and μ, such that λ represents the number of columns at the right-hand side of the diagram which have only a single square, and μ represents the number of columns in the left hand of the diagram which have two squares.

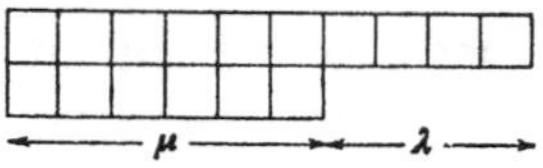

$$\lambda = 2T$$
$$\mu = \tfrac{1}{2}n - T$$

The notation (λ, μ) is used to describe these states.

Then for a system of nucleons, Young diagrams can be used to keep track of the number of states of different isospins that one gets with a given number of particles by building up these states adding up one particle at a time. We have seen that for the two-particle system there are two possible symmetries, the antisymmetric state

which would be represented by $\lambda=0$, $\mu=1$ or $(0,1)$. There is also the symmetric state

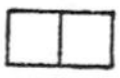

which would be represented by the quantum numbers $(2,0)$.

Let us now investigate the total number of states for the three-particle system. The first two particles can be either in the symmetry $(0,1)$ or in the symmetry $(2,0)$ and we must investigate all possible ways of adding another particle. There is only one way of adding  to 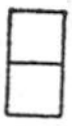: we get 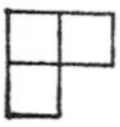 which is represented by the quantum numbers (1,1). On the other hand, there are two ways of adding  to 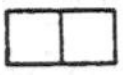. We obtain either 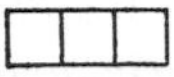, i.e. (3,0), or 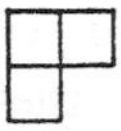, i.e. (1,1). Thus the states for a three-nucleon system include two (1,1) symmetries with $T=\frac{1}{2}$ and one (3,0) symmetry with $T=\frac{3}{2}$. To check the bookkeeping we note that there are two possible isospin states for each nucleon. There are in all $2^3=8$ possible isospin states for a three-nucleon system. One $T=\frac{3}{2}$ multiplet which has four states, and two $T=\frac{1}{2}$ multiplets each of which has two states, add up to 8 states which checks.

The total number of isospin multiplets in the system of n nucleons is obtained in an analogous manner. One adds one square  in all possible ways to the diagrams representing all the multiplets of the $n-1$-nucleon system.

The isospin formalism avoids all these bookkeeping complications by considering the nucleon to be a two-component spinor in a fictitious isospin space and then using angular momentum coupling rules. The total number of isospin multiplets for the n-nucleon system is obtained by examining the couplings of n independent spins of $\frac{1}{2}$. The first two spins can be coupled either to $T=0$ or $T=1$. The third can couple to $T=0$ to give $T=\frac{1}{2}$, or to $T=1$ to give either $T=\frac{1}{2}$ or $T=\frac{3}{2}$. We thus have three multiplets of $T=\frac{1}{2}$, $\frac{1}{2}$ and $\frac{3}{2}$ for the three-nucleon system. The fourth spin can couple to either of the $T=\frac{1}{2}$ multiplets to give $T=0$ or $T=1$, or to the $T=\frac{3}{2}$ multiplet to give $T=1$ or $T=2$. We thus have six multiplets of $T=0$, 0, 1, 1, 1 and 2 for the four-nucleon system. This can be continued indefinitely.

The isospin formalism treats permutations in nucleon systems in a simple way because the algebra of the operators which change neutrons into protons and vice versa is the same as the algebra of ordinary angular momentum. If we consider permutations in systems of protons, neutrons and Λ's we are led to the algebra of the group SU_3, as is discussed in Chapter 3. Here as in the case of isospin, the same results for the classification of states of the n-sakaton system are obtainable either by using the multiplet structure and coupling rules for SU_3, or by using permutation symmetry and Young diagrams. However, since the properties of SU_3

are not as well known as those of angular momentum, it is simpler to use permutation symmetry to classify the n-sakaton system and to use these results to determine the multiplet structure of the group SU_3. This is presented in Appendix A.

The extension of the Young diagram to the case of the Sakata model is perfectly straightforward. Since there are now three possible states for a particle, n, p and Λ, there can be a totally antisymmetric state of three particles. The Young diagrams can have three rows rather than two as in the case of nucleons. The states of the n-sakaton system can also be constructed by adding one square $\square$ to the Young diagrams representing the $n-1$-sakaton system in all possible ways. This is illustrated in Appendix A. Young diagrams can also be used to obtain rules for combining multiplets each consisting of several sakatons. The rules for adding several squares at a time to a diagram are somewhat more complicated than those for adding only one particle. These rules are not treated here and can be found in the standard literature.

CHAPTER 7

THE GROUPS SU_4, SU_6 AND SU_{12}, AN INTRODUCTION TO GROUPS OF HIGHER RANK

The methods described in Chapter 3 and 4 and in Appendices A and B for determining the structure of SU_3 multiplets can be applied as well to groups of higher rank. However, the following example illustrates the difficulties which arise with larger numbers. In SU_2, the simplest non-trivial multiplet is the doublet. The product of two doublets gives four states which reduce to two multiplets, a triplet and a singlet. For SU_3, the corresponding simplest product is that of two triplets, which reduce to a sextet and a triplet. For SU_{12} the corresponding simplest product is that of two 12-plets, which reduce to a 78-plet and a 66-plet. To describe the SU_{12} multiplet structure by diagrams analogous to those used for SU_3, an 11-dimensional space is required. Thus certain modifications in the methods for describing multiplet structure are necessary in order to treat these larger groups.

7.1. THE GROUP SU_4 AND ITS CLASSIFICATION WITH AN SU_3 SUBGROUP

The group SU_4 furnishes a convenient example in which methods usable for higher groups can be presented. The SU_4 Lie algebra can be constructed by adding a fourth particle to the three particles of the Sakata model discussed in Chapter 3. Let $a_X^\dagger$ and a_X be the operators for the creation and annihilation of this 'X'-particle. We now construct the Lie algebra of all possible bilinear products of the n, p, Λ and X operators which do not change the number of particles. From four creation operators and four annihilation operators we can make 16 possible bilinear products. One combination of these is the new baryon number, now including the number of X-particles. This operator commutes with all the others. The remaining 15 operators constitute the SU_4 Lie algebra.

The construction of the 15 operators and their commutation rules is straightforward and is left as an exercise for the reader. The algebra is of rank three since we can now find three operators which commute with one another. In addition to the operators τ_0 and N of the group SU_3 we have an additional operator which involves the number of X-particles. By analogy with the definition of the operator N in SU_3, we define a new operator Z.

$$Z = \tfrac{1}{4}(a_p^\dagger a_p + a_n^\dagger a_n + a_\Lambda^\dagger a_\Lambda - 3a_X^\dagger a_X) = \tfrac{1}{4}B + C \tag{7.1a}$$

where

$$B = a_p^\dagger a_p + a_n^\dagger a_n + a_\Lambda^\dagger a_\Lambda + a_X^\dagger a_X \tag{7.1b}$$

$$C = -a_X^\dagger a_X\,. \tag{7.1c}$$

We have defined a new quantum number, the 'charm', C, by analogy with strangeness. The X-particle has $C=-1$ while all other particles have $C=0$, just as the Λ has $S=-1$ and the other particles have $S=0$. Instead of the operators τ_0, N and Z we could have chosen any three independent linear combinations of these operators. The reason for the particular choice is analogous to the case of the Sakata model: namely to allow the subgroups SU_3 and SU_2 defined by the Sakata model and isopin respectively to be used in the classification of the SU_4 multiplets. Other possibilities are considered below.

The 15 operators of the algebra are conveniently classified as follows: The 8 operators (3.1) of the Sakata SU_3 and the operator Z all commute with Z. The 6 operators which annihilate a sakaton and create an X or vice versa do not commute with Z and change its eigenvalue by ± 1. From the expression for the operator Z we see that its eigenvalues have the form n, $n+\tfrac{1}{4}$, $n+\tfrac{1}{2}$, and $n-\tfrac{1}{4}$, where n is any integer. Since the operators of the algebra only change the eigenvalue of Z by ± 1, the eigenvalues of Z for all the states in a given SU_4 multiplet must have the same form: either n, $n+\tfrac{1}{4}$,

$n+\frac{1}{2}$, or $n-\frac{1}{4}$. Thus there are four different types of SU_4 multiplets analogous to the integral and half-integral spins of SU_2 multiplets and the three 'trialities' of SU_3 multiplets. One can show in a similar way that there are n different types of multiplets in the general case of the group SU_n.

Note that the operators like $a_p^\dagger a_X$ which change the eigenvalues of Z by one unit also change the eigenvalue of N by one-third. Thus a single SU_4 multiplet contains states having both integral and third-integral eigenvalues of N, in the same way as an SU_3 multiplet contains both integral and half-integral isospins. Just as the SU_3 multiplets consist of alternating integral and half-integral isospin multiplets as the eigenvalues of N change in steps of one unit, the SU_4 multiplets contain SU_3 multiplets whose 'triality' changes cyclically as Z increases by steps of one unit.

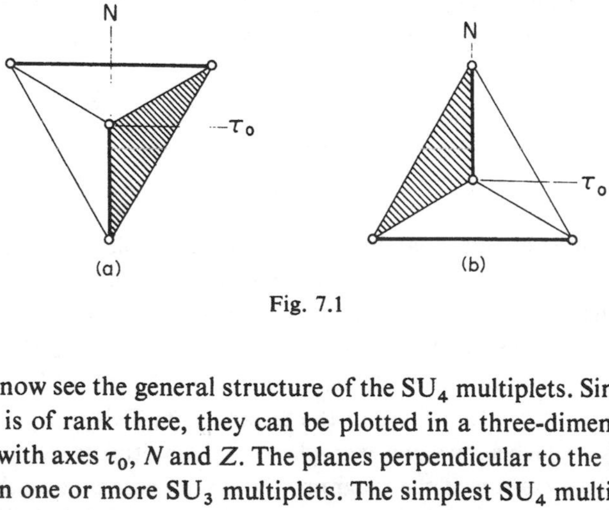

Fig. 7.1

We now see the general structure of the SU_4 multiplets. Since the group is of rank three, they can be plotted in a three-dimensional space with axes τ_0, N and Z. The planes perpendicular to the Z-axis contain one or more SU_3 multiplets. The simplest SU_4 multiplet is the quartet defined by the n, p, Λ and X particles themselves. This is a tetrahedron as shown in Fig. 7.1a and contains an SU_3 triplet (1,0) at $Z=\frac{1}{4}$ and an SU_3 singlet at $Z=-\frac{3}{4}$. The Z-axis is taken perpendicular to the paper. Instead of choosing an SU_3 subgroup corresponding to the Sakata model, we could have chosen other SU_3 subgroups corresponding to the sets of particles (pnX), (nΛX) or (pΛX). The four different SU_3 subgroups correspond to the four faces of the tetrahedron of Fig. 7.1a. The three points on each face represent a triplet in one of the SU_3 subgroups and the remaining point is the singlet of that subgroup. The corresponding antiparticle quartet is shown in Fig. 7.1b and contains an SU_3 antitriplet (0, 1) at $Z=-\frac{1}{4}$ and a singlet at $Z=+\frac{3}{4}$.

All the SU_4 multiplets can be constructed by combining quartets and using permutation symmetry in the same manner as shown for SU_3 in Appendix A. Since there are four different particle states, totally antisymmetric states of up to four particles can be constructed, but not of more than four. The totally antisymmetric four-particle state is an SU_4 singlet, with the same quantum numbers as the vacuum. The totally antisymmetric three-particle state is a quartet with the same structure as the antiparticle quartet, Fig. 7.1b. The Young diagrams characterizing the multiplets can have as many as three rows, and are characterized by three integers (λ, μ, ν) by analogy with the two integers (λ, μ) of SU_3.

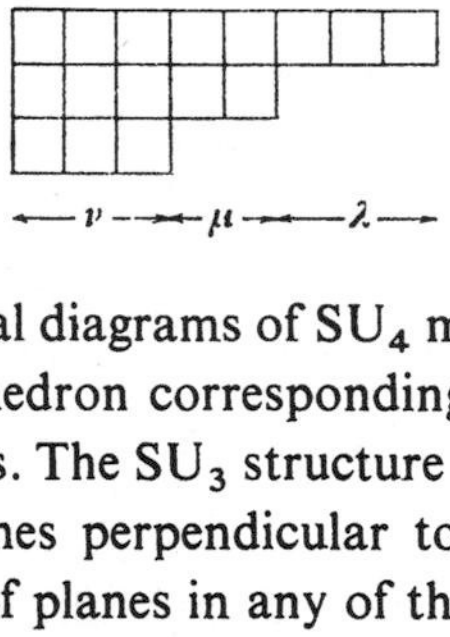

The three-dimensional diagrams of SU_4 multiplets must have the symmetry of the tetrahedron corresponding to the equivalence of the four SU_3 subgroups. The SU_3 structure of the multiplet seen in the set of parallel planes perpendicular to the Z-axis should be found in a similar set of planes in any of the four directions corresponding to the four faces of the tetrahedron. This can be seen in Fig. 7.2, which shows two examples of SU_4 multiplets. As an aid in the interpretation of the diagram, the points constituting the SU_3 multiplets of zero 'charm' (no X-particles) are labeled with the names of the particles in the baryon and meson multiplets as shown in Figs. 3.7. The X^0 is an SU_3 singlet.

For groups of high rank, the drawing of multiplet diagrams is not feasible and an alternative means for specifying the structure of a multiplet is required. A convenient means is to list the multiplets of an appropriate subgroup. For the Sakata SU_3, this can be done by listing the isospin multiplets appearing at each eigenvalue of N. For

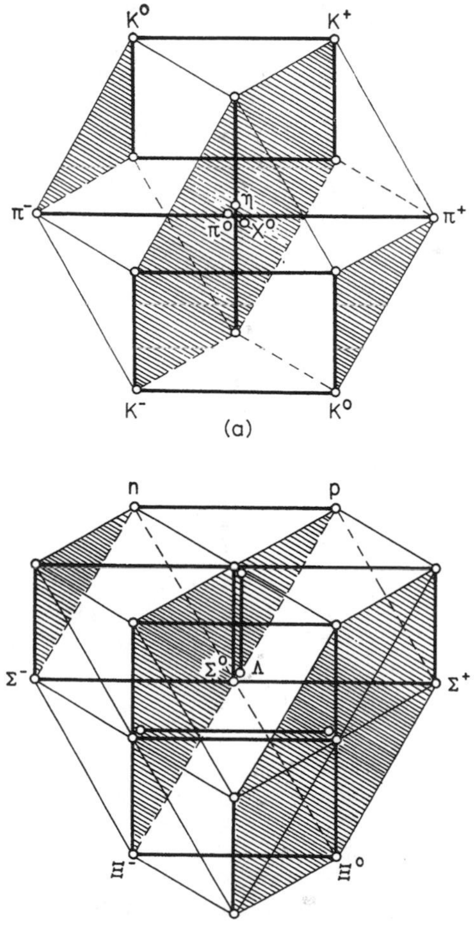

Fig. 7.2

the present version of SU_4, this can be done by listing the SU_3 multiplets appearing at each eigenvalue of Z. A common notation is to label a multiplet by the number of states which it contains;

Structure of SU_4 multiplets

SU_4 multiplet			Classification by SU_3	Z	Classification by $SU_2 \times SU_2$	X	Classification by Wigner supermultiplet $SU_2 \times SU_2$
	1	(0,0,0)	1	0	(1,1)	0	(1,1)
	4	(1,0,0)	3	$\frac{1}{4}$	(2,1)	$\frac{1}{2}$	(2,2)
			1	$-\frac{3}{4}$	(1,2)	$-\frac{1}{2}$	
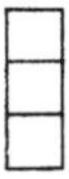	$\overline{4}$	(0,0,1)	1	$\frac{3}{4}$	(1,2)	$\frac{1}{2}$	(2,2)
			$\overline{3}$	$-\frac{1}{4}$	(2,1)	$-\frac{1}{2}$	
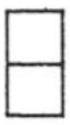	6	(0,1,0)	$\overline{3}$	$\frac{1}{2}$	(1,1)	1	(1,3)
			3	$-\frac{1}{2}$	(2,2)	0	(3,1)
					(1,1)	-1	
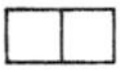	10	(2,0,0)	6	$\frac{1}{2}$	(3,1)	1	(3,3)
			3	$-\frac{1}{2}$	(2,2)	0	(1,1)
			1	$-\frac{3}{2}$	(1,3)	-1	
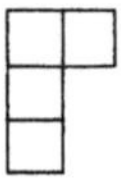	15	(1,0,1)	3	1	(2,2)	1	(3,3)
			8,1	0	(1,1)(3,1),(1,3)	0	(3,1)
			$\overline{3}$	-1	(2,2)	-1	(1,3)
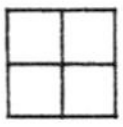	20	(0,2,0)	$\overline{6}$	1	(1,1)	2	(5,1)
			8	0	(2,2)	1	(1,5)
			6	-1	(3,3),(1,1)	0	(3,3)
					(2,2)	-1	(1,1)
					(1,1)	-2	
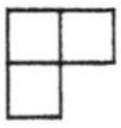	20′		8	$\frac{3}{4}$	(1,2)	$\frac{3}{2}$	(4,2)
		(1,1,0)	6,$\overline{3}$	$-\frac{1}{4}$	(2,1)(2,3)	$\frac{1}{2}$	(2,2)
			3	$-\frac{5}{4}$	(1,2),(3,2)	$-\frac{1}{2}$	(2,4)
					(2,1)	$-\frac{3}{2}$	
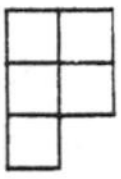	$\overline{20}$′		$\overline{3}$	$\frac{5}{4}$	(2,1)	$\frac{3}{2}$	(4,2)
			$\overline{6}$,3	$\frac{1}{4}$	(1,2),(3,2)	$\frac{1}{2}$	(2,2)
		(0,1,1)	8	$-\frac{3}{4}$	(2,1),(2,3)	$-\frac{1}{2}$	(2,4)
					(1,2)	$-\frac{3}{2}$	

SU$_4$ multiplet			*Classification by* *SU*$_3$	*Z*	*Classification by* *SU*$_2 \times SU_2$	*X*	*Classification by Wigner supermultiplet* *SU*$_2 \times SU_2$
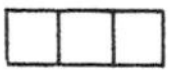	20″	(3,0,0)	10	$\frac{3}{4}$	(4,1)	$\frac{3}{2}$	(4,4)
			6	$-\frac{1}{4}$	(3,2)	$\frac{1}{2}$	(2,2)
			3	$-\frac{5}{4}$	(2,3)	$-\frac{1}{2}$	
			1	$-\frac{9}{4}$	(1,4)	$-\frac{3}{2}$	
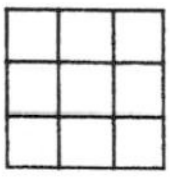	$\overline{20}''$	(0,0,3)	1	$\frac{9}{4}$	(1,4)	$\frac{3}{2}$	(4,4)
			$\overline{3}$	$\frac{5}{4}$	(2,3)	$\frac{1}{2}$	(2,2)
			$\overline{6}$	$\frac{1}{4}$	(3,2)	$-\frac{1}{2}$	
			$\overline{10}$	$-\frac{3}{4}$	(4,1)	$-\frac{3}{2}$	
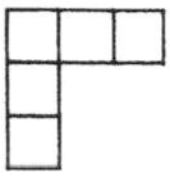	36	(2,0,1)	6	$\frac{5}{4}$	(3,2)	$\frac{3}{2}$	(4,4)
			15,3	$\frac{1}{4}$	(4,1),(2,1),(2,3)	$\frac{1}{2}$	(2,4)
			8,1	$-\frac{3}{4}$	(1,2),(1,4),(3,2)	$-\frac{1}{2}$	(4,2)
			$\overline{3}$	$-\frac{7}{4}$	(2,3)	$-\frac{3}{2}$	(2,2)
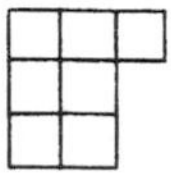	$\overline{36}$	(1,0,2)	3	$\frac{7}{4}$	(2,3)	$\frac{3}{2}$	(4,4)
			8,1	$\frac{3}{4}$	(1,2),(1,4),(3,2)	$\frac{1}{2}$	(2,4)
			$\overline{15},\overline{3}$	$-\frac{1}{4}$	(4,1),(2,1),(2,3)	$-\frac{1}{2}$	(4,2)
			$\overline{6}$	$-\frac{5}{4}$	(3,2)	$-\frac{3}{2}$	(2,2)
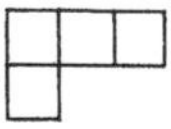	45	(2,1,0)	15	1	(3,1)	2	(5,3)
			10,8	0	(2,2),(4,2)	1	(3,5)
			6,$\overline{3}$	− 1	(3,3),(3,1),(1,3)	0	(3,3)
			3	− 2	(2,2),(2,4)	− 1	(3,1)
					(1,3)	− 2	(1,3)
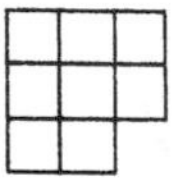	$\overline{45}$	(0,1,2)	$\overline{3}$	2	(1,3)	2	(5,3)
			$\overline{6}$,3	1	(2,2),(2,4)	1	(3,5)
			$\overline{10}$,8	0	(3,3),(3,1),(1,3)	0	(3,3)
			$\overline{15}$	− 1	(2,2),(4,2)	− 1	(3,1)
					(3,1)	2	(1,3)
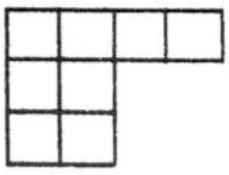	84	(2,0,2)	6	2	(3,3)	2	(5,5)
			15,3	1	(4,2),(2,2),(2,4)	1	(5,3),(3,5)
			27,8,1	0	(5,1),(3,1),(3,3) }	0	(5,1),(1,5)
			$\overline{15},\overline{3}$	− 1	(1,1),(1,3),(1,5) }		(3,3),(3,3)
			$\overline{6}$	− 2	(4,2),(2,2),(2,4)	− 1	(1,1)
					(3,3)	− 2	

e.g. the (1, 1) and (2,2) multiplets of SU_3 are denoted by 8 and 27 respectively. Conjugate multiplets are denoted by a bar; e.g. the (1,0) and (0, 1) of SU_3 by 3 and $\overline{3}$ respectively, the (3,0) and (0,3) by 10 and $\overline{10}$. In most cases relevant to particle physics there is no ambiguity in the use of a single integer to denote a multiplet, rather than the set of $n-1$ integers needed to define the Young diagram for SU_n, as there are usually not more than two conjugate multiplets having the same number of states. One exception, which occurs in SU_4, is the set of five different multiplets each of which contains 20 states. These can be denoted by 20, 20′, 20″, $\overline{20}'$ and $\overline{20}''$.

Table 7.1 lists the simplest SU_4 multiplets, giving the number of states, the Young diagram, and the SU_3 multiplets occurring at each value of Z.

7.2. THE $SU_2 \times SU_2$ MULTIPLET STRUCTURE OF SU_4

The preceding analysis of SU_4 multiplet structure has used the subgroup SU_3 corresponding to the Sakata model. We now examine other classifications obtained by the use of other subgroups. The isospin SU_2 subgroup is constructed from the operators which create or annihilate only p and n particles. Let us define another SU_2 subgroup, which we call strange spin χ, from the operators which create and annihilate only Λ and X particles.

$$\chi_+ = a^\dagger_X a_\Lambda \tag{7.2a}$$

$$\chi_- = a^\dagger_\Lambda a_X \tag{7.2b}$$

$$\chi_0 = \tfrac{1}{2}(a^\dagger_X a_X - a^\dagger_\Lambda a_\Lambda)\,. \tag{7.2c}$$

These two SU_2 subgroups commute with one another. The direct product $SU_2 \times SU_2$ of the two groups defines a subgroup of SU_4, which can be used to specify states in SU_4 multiplets. In the simple quartet multiplet, Fig. 7.1, the p and n form an isospin doublet which is a strange-spin singlet. We denote this by (2, 1) where the first number labels the isospin multiplet and the second the strange spin. This is a common notation for labeling multiplets in a group which is the direct product of two smaller groups. The Λ and X particles then form a (1,2) multiplet of $SU_2 \times SU_2$; i.e. an isospin singlet and strange-spin doublet.

The strange-spin multiplets are exhibited simply in the SU_4 multiplet diagrams of Figs. 7.1 and 7.2. They are shown as heavy vertical lines, while isospin multiplets are shown as heavy horizontal lines.

The operators N and Z which were used with SU_3 are not convenient for defining quantum numbers for this $SU_2 \times SU_2$ classification. Instead we choose two different linear combinations of these operators, the strange-spin operator χ_0 and an operator X defined as half the difference between the number of ordinary (pn) particles and the number of strange and charmed particles (ΛX).

$$X = \tfrac{1}{2}(a_p^\dagger a_p + a_n^\dagger a_n - a_X^\dagger a_X - a_\Lambda^\dagger a_\Lambda)\,. \tag{7.3}$$

The quartet multiplet thus has a (2,1) multiplet of $SU_2 \times SU_2$ at $X = +\frac{1}{2}$ and a (1,2) multiplet of $SU_2 \times SU_2$ at $X = -\frac{1}{2}$.

The structure of SU_4 multiplets expressed in terms of this $SU_2 \times SU_2$ subgroup is easily obtained by combining quartets and using permutation symmetry by analogy to the procedure used for SU_3 in Appendix A. This classification is also listed in Table 7.1, giving the $SU_2 \times SU_2$ multiplets occurring at each value of X.

7.3. THE WIGNER SUPERMULTIPLET SU_4

The SU_4 group was first introduced to nuclear physics in the Wigner supermultiplet description of light nuclei. Here the multiplet structure was described by an $SU_2 \times SU_2$ subgroup different from the $SU_2 \times SU_2$ subgroup which we have just considered.

The Wigner supermultiplet theory considered the equivalence of the four spin and isospin states of the nucleon which we can denote by p↑, p↓, n↑, n↓. Let us define our algebra using the creation and annihilation operators for these four states rather than for n, p, Λ, X. This does change any algebraic properties, but gives a different physical point of view. For example, the four states p↑, p↓, n↑, n↓, can be divided into two pairs in various ways with some physical significance. However, there is no physically meaningful way to divide them into a triplet and a singlet. Thus no SU_3 subgroup arises naturally in this case. There are several $SU_2 \times SU_2$ subgroups analogous to isospin and strange spin in the previous example. One is the direct product of the ordinary spin of the protons and the ordinary spin of the neutrons. That this leads to the same multiplet structure can be seen by examining the structure of the fundamental 4-multiplet. It again contains two multiplets of $SU_2 \times SU_2$; namely a (2, 1) and (1,2). Since all larger multiplets can be constructed from quartets they also have the same structure.

One can also define other equivalent $SU_2 \times SU_2$ subgroups by considering, for example, the isospin of the particles having ordinary spin 'up' and the isospin of the particles having ordinary spin 'down'. However, it is also possible to define a completely different $SU_2 \times SU_2$ subgroup which is more natural for this problem; namely the direct product of isospin and ordinary spin. That this leads to a different classification of the multiplet structure is immediately evident from examination of the fundamental quartet. The four states of the nucleon are a doublet in both isospin and ordinary spin and therefore constitute a single (2,2) multiplet of $SU_2 \times SU_2$, instead of containing a (2, 1) and a (1,2) as in the previous case.

The structures of the simplest SU_4 multiplets are easily obtained in the Wigner supermultiplet classification. The use of permutation symmetry is simpler than in the previous $SU_2 \times SU_2$ classification because the fundamental quartet contains only a single multiplet of $SU_2 \times SU_2$. The larger multiplets can be constructed by combining a number of quartets and using permutation symmetry as is done for SU_3 in Appendix A. The $SU_2 \times SU_2$ states which can appear in a given SU_4 multiplet are determined by the permutation symmetry.

As an example, consider the combination of two quartets. The combination of two (2,2) multiplets of $SU_2 \times SU_2$ gives

$$(2,2) \times (2,2) = \underbrace{(3,3)+(1,1)}_{\text{S}}+\underbrace{(3,1)+(1,3)}_{\text{A}} \tag{7.4a}$$

where S and A denote the symmetric and antisymmetric combinations. The (3, 3) is symmetric in both isospin and ordinary spin; the (1, 1) is antisymmetric in both. These are thus both symmetric in

$SU_2 \times SU_2$ and therefore also in SU_4. The (3,1) and (1,3), being symmetric in one SU_2 and antisymmetric in the other, are antisymmetric in $SU_2 \times SU_2$ and in SU_4. We thus obtain the SU_4 relations

$$4 \times 4 = 10 + 6 \tag{7.4b}$$

where

$$10 = (3,3) + (1,1) \tag{7.4c}$$

$$6 = (3,1) + (1,3)\,. \tag{7.4d}$$

The isospin–ordinary-spin structure of SU_4 multiplets is also given in Table 7.1.

Let us examine the choice of three SU_4 operators to be diagonal in this rank-three algebra for the case of the Wigner supermultiplet classification. The operators N and Z used with SU_3 are not suitable as they suggest that one of the four states is different from the other three. We can use τ_0 and S_z, the z-component of the ordinary spin. For the third operator, we choose the operator

$$W = \tfrac{1}{2}(a^\dagger_{p\uparrow} a_{p\uparrow} + a^\dagger_{n\downarrow} a_{n\downarrow} - a^\dagger_{p\downarrow} a_{p\downarrow} - a^\dagger_{n\uparrow} a_{n\uparrow})\,. \tag{7.5}$$

For the states of the fundamental quartet, W is not independent of τ_0 and S_z and is equal to $2\tau_0 S_z$. This tends to cause confusion. Since the two quantum numbers τ_0 and S_z are already sufficient to specify a member of the quartet completely, W is redundant in the quartet and must be a function of the other two. However, this is not true in a general SU_4 multiplet; otherwise W would not be needed and the group would not be of rank three. In fact, even in other multiplets which are still so small that W is redundant, it is not the *same* function $2\tau_0 S_z$ as in the 4. Consider for example the case of the antiquartet $\bar{4}$, which contains the antinucleons. For this multiplet the operator W defined by equation (7.5) is not equal to the product $2\tau_0 S_z$ but to $-2\tau_0 S_z$.

The operator W does not commute with either of the Casimir operators T^2 and S^2 of the SU_2 subgroups of isospin and ordinary spin. Therefore the operator W cannot be used to classify the states in SU_4 multiplets, if these states are chosen to be eigenstates of the total isospin and total spin. There is no third operator in the Lie algebra which commutes with τ_0, S_z, T^2 and S^2. This is analogous to the classification of SU_3 multiplets using the R_3 subgroup discussed in § 4.2. It is therefore impossible to display SU_4 multiplets on a three-dimensional plot if the states are to be classified in isospin and ordinary spin multiplets. There is no quantum number to label the third axis.

In the Wigner supermultiplet classification the number of quantum numbers available is insufficient to give a complete specification of the states within the multiplet. This can be seen by comparing the number of quantum numbers available with those used in the classification with the SU_3 subgroup. In the SU_3 case, three quantum numbers are defined by Casimir operators of SU_4 subgroups: the two Casimir operators of SU_3 and the one of the isospin SU_2. In the Wigner supermultiplet case there are only the two Casimir operators of isospin and ordinary spin. Furthermore the quantum number W has also been lost. However, the full set of quantum numbers is not needed to classify the states in the smaller multiplets which are the ones usually used in physical problems. For example, in Table 7.1 we see that the difficulty first arises only in the 84 multiplet which contains two (3, 3) multiplets of $SU_2 \times SU_2$.

The two different ways of obtaining $SU_2 \times SU_2$ subgroups of SU_4 occur as well in larger unitary groups. That the two subgroups are isomorphic is an accident for the case of SU_4 because $4=2+2=2 \times 2$. The isospin–strange-spin classification is obtained by separating the fundamental quartet into two completely independent pairs and defining an SU_2 group for each one. Here $4=2+2$ is relevant. The isospin–ordinary-spin classification is obtained by separating the fundamental into two doublets in *two different ways*. Here $4=2 \times 2$ is relevant.

7.4. THE GROUP SU_6

The Lie algebra SU_6 arises naturally in an attempt to combine the Sakata model and the Wigner supermultiplet. Consider the six states obtained from both spin states for each of the three sakatons. The algebra of all bilinear products of six creation and six annihilation operators leads to 36 operators. One linear combination is the baryon number; the remaining 35 form the Lie algebra of SU_6. The group is of rank five. There are six independent commuting number operators, and five independent linear combinations remain after the removal of the baryon number. Plots of diagrams of these operators and of the multiplets in a five-dimensional space are not only impractical but are not even desirable for most purposes.

A complete classification of the states of SU_6 multiplets in which all the number operators are diagonal can be obtained by using the chain of subgroups SU_5, SU_4, SU_3 and SU_2, in a manner directly analogous to that used for SU_4 in § 7.1 and SU_3 in Chapter 3. Such classification is not of physical interest for this model as there is no SU_5 subgroup which has any direct physical interpretation. Classification of the SU_6 multiplets can also be obtained which uses subgroups of SU_6 of physical interest, such as the SU_3 group of the Sakata model and the SU_4 group of the Wigner supermultiplet. These do not commute with all the number operators and lead to a multiplet structure which cannot be plotted on a five-dimensional diagram.

We consider here two classifications of the multiplets of SU_6, using subgroups which are direct products. These are directly analogous to the two $SU_2 \times SU_2$ subgroups used in the treatment of SU_4. One case is the direct product of the Wigner supermultiplet SU_4 and the SU_2 describing the spin of the Λ particles. The other is the direct product of the Sakata SU_3 and the SU_2 of ordinary spin. These are directly analogous to the isospin–strange-spin and isospin–ordinary-spin classifications we considered for SU_4. The two groups correspond to the decompositions $6=4+2$ and $6=3 \times 2$ respectively.

Consider first the classification according to the subgroup $SU_4 \times SU_2$. The Wigner supermultiplet SU_4 acts only on p and n; the SU_2 is the Λ spin and acts only on the spin states of the Λ. The quantum number N, defined as in eq. (3.1) for the Sakata model (with the implied sum over spin states) is a good quantum number for this classification in addition to the quantum numbers of $SU_4 \times SU_2$. Thus the structure of the SU_6 multiplets is described by giving the $SU_4 \times SU_2$ multiplets appearing at each value of N. This is directly analogous to the isospin–strange-spin classification for SU_4.

The structure of the fundamental sextet of SU_6 can be seen by examining the six states of the sakaton. The four states of the p and n form a quartet of SU_4 and singlet of SU_2, denoted by (4, 1), at $N=\frac{1}{3}$. The remaining two states of the Λ constitute a (1, 2) at $N=-\frac{2}{3}$. The SU_6 multiplets can now be built up by combining sakatons in a manner directly analogous to that described in detail in Appendix A for SU_3 and used in the preceding example for SU_4.

Let us now examine the classification using the subgroup $SU_3 \times SU_2$. Here SU_3 operates entirely in the space of isospin and strangeness while SU_2 operates only in spin space. The sextet sakaton multiplet is a single multiplet of $SU_3 \times SU_2$, a triplet in SU_3 and a doublet in SU_2, denoted by (3,2). This is analogous to the Wigner supermultiplet classification of SU_4 where the fundamental quartet was a single multiplet (2, 2) in $SU_2 \times SU_2$. The construction of the higher multiplets by combining sakatons follows very closely the corresponding procedure for SU_4. The permutation symmetry of the particular SU_6 multiplet constructed from sakatons determines the allowed permutation symmetries of the $SU_3 \times SU_2$ multiplets which it can contain. By direct analogy with eq. (7.4) for SU_4, we have

$$(3,2)\times(3,2) = \underbrace{(6,3)+(\bar{3},1)}_{\text{S}} + \underbrace{(\bar{3},3)+(6,1)}_{\text{A}}, \tag{7.6a}$$

$$6\times 6 = 21+15, \tag{7.6b}$$

$$21 = (6,3)+(\bar{3},1), \tag{7.6c}$$

$$15 = (\bar{3},3)+(6,1). \tag{7.6d}$$

Some of the SU_6 multiplets and their structures are listed in Table 7.2. Both the $SU_4 \times SU_2$ and the $SU_3 \times SU_2$ classifications are given.

The group SU_6 has been applied to elementary particles in the

Structure of SU_6 multiplets

SU_6 multiplet		*Classification by* $SU_4 \times SU_2$	N	*Classification by* $SU_3 \times SU_2$
	1	(1,1)	0	(1,1)
	6	(4,1)	$\frac{1}{3}$	(3,2)
		(1,2)	$-\frac{2}{3}$	
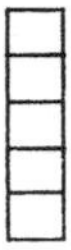	$\overline{6}$	(1,2)	$\frac{2}{3}$	$(\overline{3},2)$
		$(\overline{4},1)$	$-\frac{1}{3}$	
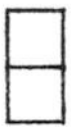	15	(6,1)	$\frac{2}{3}$	$(\overline{3},3)$
		(4,2)	$-\frac{1}{3}$	(6,1)
		(1,1)	$-\frac{4}{3}$	
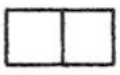	21	(10,1)	$\frac{2}{3}$	(6,3)
		(4,2)	$-\frac{1}{3}$	$(\overline{3},1)$
		(1,3)	$-\frac{4}{3}$	
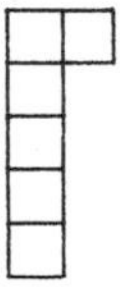	35	(4,2)	1	(8,3)
		(15,1)(1,3)(1,1)	0	(8,1)
		$(\overline{4},2)$	-1	(1,3)
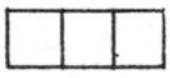	56	(20″,1)	1	(10,4)
		(10,2)	0	(8,2)
		(4,3)	-1	
		(1,4)	-2	
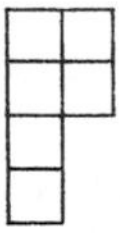	189	(6,1)	2	(27,1)
		$(\overline{20}',2)$(4,2)	1	(10,3)$(\overline{10},3)$
		(20,1)(15,3)(15,1)(1,1)	0	(8,5)(8,3)(8,3)(8,1)
		$(\overline{4},2)$(20′,2)	-1	(1,5)(1,1)
		$(\overline{6},1)$	-2	
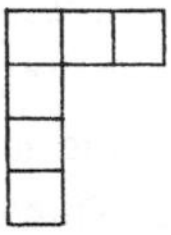	280	(10,1)	2	(27,3)
		(36,2)(4,2)	1	(10,5)(10,3)(10,1)
		(45,1)(15,3)(15,1)(1,3)	0	$(\overline{10},1)$
		(20′,2)$(\overline{4},4)(\overline{4},2)$	-1	(8,5)(8,3)(8,3)(8,1)
		(6,3)	-2	(1,3)
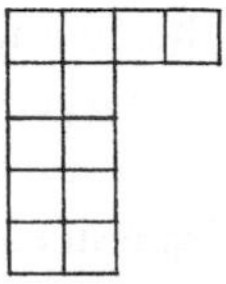	405	(10,3)	2	(27,5)(27,3)(27,1)
		(36,2)(4,4)(4,2)	1	(10,3)$(\overline{10},3)$
		(84,1)(15,3)(15,1)(1,5) (1,3)(1,1)	0	(8,5)(8,3)(8,3)(8,1)
		$(\overline{36},2)(\overline{4},4)(\overline{4},2)$	-1	(1,5)(1,1)
		$(\overline{10},3)$	-2	

octet model, not in the Sakata model which we have used to determine the algebraic properties of the group. Let us now examine the classification of mesons and baryons in SU_6 multiplets using the interpretation of the SU_3 subgroup given by the octet model. We immediately find the remarkable fact that all of the four SU_3 multiplets appearing in Fig. B.1, Appendix B, fit neatly into two SU_6 multiplets and have just the right values for the spins as well as the SU_3 quantum numbers. The baryon octet and decuplet fit into an SU_6 56 while the vector and pseudoscalar mesons fit into an SU_6 35.

TABLE 7.3

SU_6 classifications of mesons and baryons

	$SU_4 \times SU_2$	N	*Particles*	$SU_3 \times SU_2$	*Particles*
35	(4,2)	1	(K, K^*)	(8,3)	$(\varrho, \omega_8, K^*, \overline{K^*})$
	(15,1)	0	(ϱ, π, ω)	(8,1)	$(\pi, \eta, K, \overline{K})$
	(1,3)	0	(φ)	(1,3)	(ω_1)
	(1,1)	0	(η)		
	$(\overline{4},2)$	-1	$(\overline{K}, \overline{K}^*)$		
56	(20″,1)	1	(N, N^*)	(10,4)	$(N^*, Y^*, \Xi^*, \Omega)$
	(10,2)	0	(Σ, Λ, Y^*)	(8,2)	$(N, \Lambda, \Sigma, \Xi)$
	(4,3)	-1	(Ξ, Ξ^*)		
	(1,4)	-2	(Ω)		

The classification of the particles as they appear in these two supermultiplets is presented in Table 7.3, giving both the $SU_3 \times SU_2$ and the $SU_4 \times SU_2$ classifications. For these simple multiplets the same states appear as eigenstates in both classifications except for the case of ω and φ. This is because all states except ω and φ are uniquely specified by their spin, isospin and hypercharge, which are good quantum numbers in both the $SU_4 \times SU_2$ and the $SU_3 \times SU_2$ classifications. The ω and φ are the only example in the 56 and 35 of two states having the same spin, isospin and hypercharge. The SU_3 eigenstates, ω_1 and ω_8 naturally appear in $SU_3 \times SU_2$ classification. The two states appearing in the $SU_4 \times SU_2$ classification are linear combinations of ω_1 and ω_8 which seem to be very close to the states describing the experimentally observed ω and φ particles. These are therefore simply labeled in Table 7.3 as ω and φ.

The determination of the proper singlet and octet mixing for the ω and φ, which was completely extraneous in SU_3, arises naturally in SU_6. This was one of the early successes of this theory. One characteristic feature of the mixing, namely that the decay $\varphi \to \varrho\pi$ is forbidden, was mysterious in SU_3 and follows naturally from SU_6. From the $SU_4 \times SU_2$ classification in Table 7.3 we see that the φ belongs to a $(1,3)$ and is a triplet in the SU_2 subgroup which we call Λ-spin, while the ϱ and π belong to a $(15,1)$ and are both singlets in this SU_2 subgroup. If the decay conserves SU_6 it conserves this SU_2 subgroup and the φ which is a triplet cannot decay into ϱ and π which are both singlets.

In this description of elementary particles the baryons are classified in the 56 multiplet which is constructed in the Sakata model out of three sakatons in a totally symmetric state. It is as if there existed a fundamental SU_3 triplet of spin $\frac{1}{2}$ from which all particles were constructed. Such triplets have been called quarks and their existence is still an open question at the time of writing of this chapter.

7.5. THE GROUP SU_{12}

The SU_{12} algebra has also recently become of interest in the discussion of symmetries of elementary particles. This algebra has $144-1=143$ operators and is of rank 11. A simple way to determine the relevant properties of the multiplet structure is to add another two-valued degree of freedom to the sakaton or quark discussed in SU_6. This might correspond to the description of the spin degree of freedom by a four-component Dirac spinor instead of a two-component Pauli spinor. For purposes of the algebra we assume the existence of an additional SU_2 group which we call ϱ-spin, corresponding to the three ϱ matrices of Dirac. The fundamental multiplet of SU_{12}, from which we construct all the other multiplets, is a sakaton of 12 states which contains an SU_6 sextet with 'ϱ-spin up' and a sextet with 'ϱ-spin down'. If one wishes to have a better physical picture of the ϱ-spin one can replace the up and down ϱ-spin states by positive and negative energy states or by large and small components in a Dirac spinor.

The structure of the SU_{12} multiplets can be described in a variety of ways using subgroups. We consider three possibilities: $SU_6 \times SU_6$, $SU_6 \times SU_2$ and $SU_4 \times SU_3$.

In the $SU_6 \times SU_6$ classification we have separate SU_6 groups for the particles having ϱ-spin up and those having ϱ-spin down. The zero component of the ϱ-spin, ϱ_0, commutes with both of the SU_6 groups and provides an additional additive quantum number. The SU_{12} multiplets are thus described by giving the particular $SU_6 \times SU_6$ multiplets which occur at each value of ϱ_0. In this classification the fundamental 12 multiplet breaks up into two sextets, a $(6, 1)$ of $SU_6 \times SU_6$ at $\varrho_0 = +\frac{1}{2}$ and a $(1, 6)$ at $\varrho_0 = -\frac{1}{2}$. The conjugate $\overline{12}$ has a $(1, \overline{6})$ at $\varrho_0 = \frac{1}{2}$ and a $(\overline{6}, 1)$ at $\varrho_0 = -\frac{1}{2}$. Using this classification for the 12 multiplet it is a simple but tedious procedure to construct the other multiplets. The numbers are larger than those appearing in SU_6 but the procedure is straightforward.

The classification of multiplets of SU_{12} using a subgroup which is the direct product of the ordinary SU_6 group and the ϱ-spin SU_2 group is directly analogous to the classification of SU_6 using the subgroup $SU_3 \times SU_2$. Here the fundamental 12 multiplet appears as a single $(6, 2)$ multiplet in $SU_6 \times SU_2$. Thus the construction of the SU_{12} multiplets is facilitated by noting that the permutation symmetry of the particular SU_{12} multiplet, as indicated by its construction from the fundamental 12 multiplet, determines completely the allowed permutation symmetries for the constituent $SU_6 \times SU_2$ multiplets.

Another possible classification is the direct product of the SU_4 subgroup generated by the combination of ϱ-spin and ordinary spin, and the ordinary SU_3 subgroup. The SU_4 subgroup for this case is just the Dirac algebra for the case where the ordinary spin and ϱ-spin represent the transformations of four-component Dirac spinors. Here the fundamental 12 multiplet contains a single $(4, 3)$ multiplet of $SU_4 \times SU_3$. The construction of higher multiplets from the 12 multiplet using permutation symmetry is straightforward.

We present two examples, which correspond to the 35 and 56 of SU_6.

The multiplet in which the 143 operators of the algebra are classified is constructed by combining a fundamental 12 with its conjugate $\overline{12}$ and removing a singlet. The structure of the 143 is immediately seen to be

1. $SU_6 \times SU_6$: $(6,\bar{6})$ at $\varrho_0 = 1$;
   $(35,1)(1,35)(1,1)$ at $\varrho_0 = 0$;
   $(\bar{6},6)$ at $\varrho_0 = -1$.
2. $SU_6 \times SU_2$: $(6,2) \times (\bar{6},2) - (1,1) = (35,3) + (35,1) + (1,3)$.
3. $SU_4 \times SU_3$: $(4,3) \times (\bar{4},\bar{3}) - (1,1) = (15,8) + (1,8) + (15,1)$.

The totally symmetric multiplet constructed from three 12's has $12 \times 13 \times 14/1 \times 2 \times 3 = 364$ states. The structure is seen to be

1. $SU_6 \times SU_6$: $(56,1)$ at $\varrho_0 = \frac{3}{2}$;
   $(21,6)$ at $\varrho_0 = \frac{1}{2}$;
   $(6,21)$ at $\varrho_0 = -\frac{1}{2}$;
   $(1,56)$ at $\varrho_0 = -\frac{3}{2}$.
2. $SU_6 \times SU_2$: $[(6,2) \times (6,2) \times (6,2)]$ sym. $=$
   $(56,4) + (70,2)$
3. $SU_4 \times SU_3$: $[(4,3) \times (4,3) \times (4,3)]$ sym. $=$
   $(20'',10) + (20',8) + (\bar{4},1)$.

APPENDIX A

CONSTRUCTION OF THE SU_3 MULTIPLETS BY COMBINING SAKATON TRIPLETS

In this appendix we shall see how the multiplets of the SU_3 group can be constructed by combining the simple sakaton triplets of Fig. 3.2a. We shall examine the states of the n-sakaton system and group them into SU_3 multiplets characterized by the values of the quantum numbers λ and μ. In this way we determine the structure of the multiplets, their shape as indicated by Fig. 3.4 and the multiplicity of states at each point of the diagram, i.e. how many isospin multiplets there are for a given value of strangeness, or N.

In considering the multiplets arising in the n-sakaton system, it is convenient to examine the permutation symmetry of the different states. The operators (3.1) of the Lie algebra act symmetrically on all the particles in an n-sakaton system and therefore cannot change the permutation symmetry of the state. For example, these operators acting on a symmetric 2-sakaton state can only give other symmetric 2-sakaton states. They cannot mix symmetric and antisymmetric states. Thus all the states of the n-sakaton system which belong to a given SU_3 multiplet must have the same permutation symmetry and, conversely, two states having different permutation symmetries must belong in different SU_3 multiplets.

The use of Young diagrams is convenient in discussing permutation symmetry, but is not necessary for the treatment given below. The reader who is unfamiliar with them can safely skip all references to Young diagrams without losing any essential points. The reader who is familiar with Young diagrams should find that they are helpful but not essential.

The quantum numbers λ and μ are introduced in two different ways: (1) as labels for a multiplet defined by the shape of the multiplet diagram (Fig. 3.4) and (2) as labels for permutation symmetry defined by the shape of a Young diagram. We shall see below that the two definitions are equivalent if the Young diagram refers to the permutation symmetry of the n-sakaton state from which the multiplet is constructed.

In discussing permutation symmetry for SU_3, one can think of a 'generalized isospin'. A many-sakaton wave function is considered to be the product of a space-spin part and an SU_3 part, by analogy with the usual separation of a many-nucleon wave function into a space-spin part and an isospin (SU_2) part. In the Sakata model the 'generalized Pauli principle' would require that the overall many-sakaton wave function be antisymmetric in space-spin and SU_3. *Permutation symmetry discussed below always refers to the SU_3 part of the many-sakaton wave function.*

We first examine the two-sakaton system. Since there are three possible states for each sakaton we see that there are nine possible states for the two-sakaton system. The values of strangeness occurring are zero for two nucleons, -1 for a nucleon and Λ, and -2 for

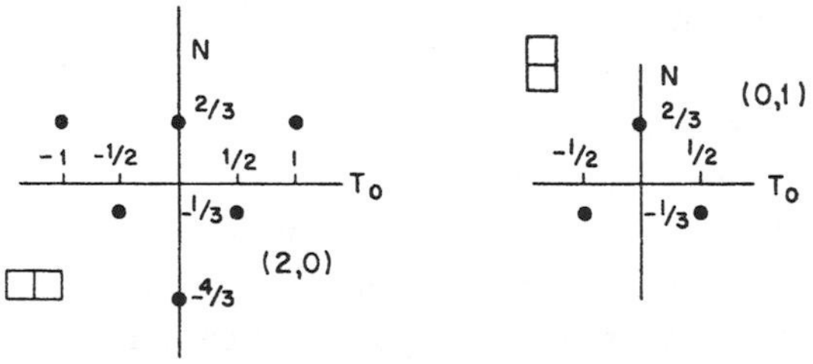

Fig. A.1

two Λ's. Since the baryon number is 2, the corresponding eigenvalues of N are $\frac{2}{3}$, $-\frac{1}{3}$ and $-\frac{4}{3}$. The $S=0$ two-nucleon combination contributes four states, an isospin triplet, $T=1$, and an isospin singlet, $T=0$. A nucleon and a Λ give an isospin doublet, $T=\frac{1}{2}$, $S=-1$, and similarly for a Λ and a nucleon. Two Λ's give an $S=-2$, $T=0$ singlet.

These nine states can be conveniently divided into two SU_3 multiplets by making use of permutation symmetry. The symmetric and antisymmetric combinations of the two-sakaton states form two distinct multiplets, as shown in Fig. A.1. The two multiplets are conveniently labeled by the appropriate Young diagram and the quantum numbers (λ, μ). The construction of the two diagrams of Fig. A.1 is easily achieved as follows: We note that the $S=0$ two-nucleon states divide into the symmetric, $T=1$, and the antisymmetric, $T=0$. The two-Λ singlet with $S=-2$ is clearly symmetric. The two doublets of the Λ–N system can be arranged to form two linear combinations, one of which is symmetric and the other antisymmetric.

Before continuing further in the construction of multiplets, we note that there exists a three-sakaton state which is totally antisymmetric since there are three different kinds of sakatons rather than two as for nucleons. Such a three-sakaton state would be denoted by the Young diagram

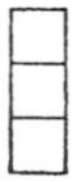

Young diagrams for many-sakaton systems would therefore have the general form

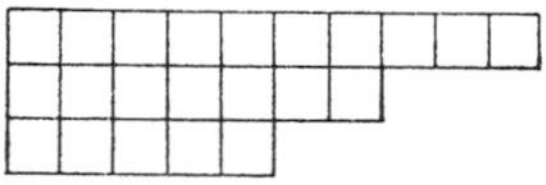

including some columns having three squares as well as columns having two squares and one square. However, we note that the completely antisymmetric three-sakaton state must be an isosinglet with baryon number 3, strangeness -1, and therefore has the quantum numbers $N=0$, $T=0$, $T_0=0$. All the operators (3.1) except the baryon number give zero when operating on this state. The antisymmetric three-sakaton state thus has the same SU_3 quantum numbers as the vacuum. If we are only interested in the structure of the SU_3 multiplet, we can disregard the columns having three squares in any Young diagram. These simply add three units to the baryon number without changing any other quantum numbers. We can therefore find the structure of all possible multiplets by examining only those Young diagrams consisting of columns containing only 1 or 2 squares. The quantum numbers (λ, μ) are therefore sufficient to describe these diagrams and therefore each SU_3 multiplet is characterized by the two appropriate quantum numbers (λ, μ).

The totally antisymmetric three-sakaton state is an SU_3 multiplet which is a singlet having the same quantum numbers as the vacuum. This singlet is represented by the quantum numbers $(\lambda, \mu) = (0, 0)$. The multiplet (0, 1) representing the antisymmetric two-sakaton system also represents the antisakaton triplet, $\bar{\Lambda}$, $\bar{p}$, and $\bar{n}$. This is not surprising since the antisymmetric two-sakaton system is formed by removing a sakaton from the antisymmetric (0, 0) three-sakaton state which has the same quantum numbers as the vacuum.

We now consider all the multiplets arising in the three-sakaton system. Since there are three states for each sakaton there are in all 27 possible states for the three-sakaton system. Let us now divide these 27 states into SU_3 multiplets. One state is clearly the totally antisymmetric three-sakaton state mentioned above which stands by itself in a (0,0) singlet. Another multiplet is formed by all the states which are totally symmetric.

The structure of the totally symmetric three-sakaton multiplet is easily obtained. The maximum value of strangeness is zero and this corresponds to $N=1$, since $B=3$. The $N=1$ part of the multiplet consists of a totally symmetric 3-nucleon state which is therefore an isospin quartet with $T=\frac{3}{2}$. The $N=0$ $(S=-1)$ part of this multiplet consists of states of one Λ and two nucleons which are totally symmetric and therefore constitute an isospin triplet with $T=1$. The $N=-1$ $(S=-2)$ part of the multiplet consists of states of a single nucleon and two Λ's and must therefore be an isospin doublet with $T=\frac{1}{2}$. The $N=-2$ part of the multiplet is a three-Λ state which is an isospin singlet with $T=0$. The totally symmetric three-sakaton states thus form a (3,0) multiplet as is shown in Fig. 3.6a. It is also evident that the totally symmetric three-sakaton state is represented by the Young diagram

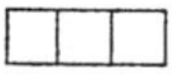

which has the appropriate quantum numbers (3,0). The (3,0) multiplet has 10 states and is therefore called a decuplet.

We now see that the (0,0) singlet and the (3,0) decuplet have used up $1+10=11$ of the 27 states of the three-sakaton system. Sixteen states remain to be accounted for.

The analysis of the remaining 16 states is facilitated by noting that we can consider the three-sakaton system as constructed by adding a sakaton to a two-sakaton system. The two-sakaton system can be either in the symmetric (2,0) multiplet with 6 states or the antisymmetric (0,1) multiplet with 3 states. Adding a sakaton to the symmetric (2,0) two-sakaton multiplet gives a total of $6\times3=18$ possible states. Ten of these states must constitute the totally symmetric (3,0) three-sakaton multiplet, leaving 8 states to be classified. If we add a sakaton to the antisymmetric (0,1) two-sakaton multiplet there is a total of $3\times3=9$ states. One of these states is the totally antisymmetric (0,0) multiplet, and we are again left with 8 states to be classified. These two sets of 8 states add up to the 16 mentioned above. One might suspect that these 2 sets of 8 states would both correspond to multiplets of the same kind. This suspicion is supported by examining the quantum numbers of the states which remain to be classified. We find that both sets of 8 states have the same quantum numbers; namely, an isospin doublet $T=\frac{1}{2}$ with $N=+1$, an isospin triplet ($T=1$) and an isospin singlet ($T=0$) with $N=0$, and an isospin doublet ($T=\frac{1}{2}$) at $N=-1$. Plotting these states in a diagram one finds the structure of the (1,1) octet shown in Fig. 3.3.

Let us now examine the octet of three-sakaton states in detail and verify that they indeed form a single SU_3 multiplet, and in particular that the isospin singlet $T=0$ at $N=0$ belongs in the (1,1) multiplet and is not another separate (0,0) multiplet. We note that the state of maximum strangeness in this multiplet is the $S=0$, $N=+1$ three-nucleon state. This state cannot be totally symmetric ($T=\frac{3}{2}$) because the totally symmetric state belongs in the totally symmetric (3,0) three-sakaton multiplet. The only other possible symmetry for a three-nucleon system has $T=\frac{1}{2}$. This symmetry is represented by the (1,1) Young diagram

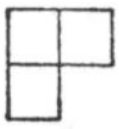

and again illustrates the equivalence of (λ, μ) classifications by the shape of the multiplet and by the Young diagram. The minimum value of N occurring in this multiplet is $N=-1$ ($S=-2$), corresponding to two Λ's and a nucleon, since the three-Λ system must be totally symmetric and can only appear in the totally symmetric (3,0) multiplet. This $N=-1$ part of the multiplet is clearly an isospin doublet since a one-nucleon, two-Λ state cannot be anything else.

The $N=0$ ($S=-1$) part of the multiplet now remains to be investigated. This is a two-nucleon, one-Λ system and can be either an isospin triplet ($T=1$) or an isospin singlet ($T=0$). That both the triplet and the singlet occur in the (1,1) multiplet can be seen as follows. The states of the $N=0$ part of the multiplet can be obtained from the states of the $N=1$, two-nucleon isospin doublet by the operation of the operators C_+ or C_- which change a nucleon into a Λ. The three-nucleon doublet can be considered to be composed of two nucleons which are coupled to total isospin 0, with an additional odd nucleon giving the isospin $\frac{1}{2}$. We see that if we change the odd nucleon into a Λ we have effectively killed off its isospin and are left with the isospin of the $T=0$ nucleon pair. This state is an isospin singlet. On the other hand, if we change one of the two nucleons in the pair to a Λ, we kill off its isospin thereby leaving the isospin $\frac{1}{2}$ of its partner nucleon free to couple with the odd nucleon to give $T=1$, a triplet, or $T=0$, a singlet. We see therefore that the $N=0$ part of the (1,1) SU_3 multiplet contains two kinds of isospin multiplets, a triplet ($T=1$) and a singlet ($T=0$). Since we obtained the isospin singlet in two ways, we might ask whether there might be two isospin singlets in the (1,1) multiplet; i.e. whether one can form two linearly independent $T=0$ states by operating on the $N=1$ doublet with the C-operators which change a nucleon into a Λ. That this is *not* the case is evident from counting states. We already use up all the eight states with only a single $T=0$ state. The structure of the $(1,1)$ multiplet is illustrated in Fig. 3.3b.

In our analysis of all the multiplets arising in one-, two-, and three-sakaton systems we have also incidentally found the following relations for combining multiplets.

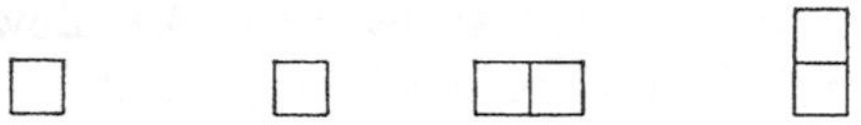

$$\begin{array}{ccccccc} (1,0) & \times & (1,0) & \rightarrow & (2,0) & + & (0,1) \\ 3 & \times & 3 & = & 6 & + & 3 \end{array} \qquad (A.1a)$$

$$\begin{array}{ccccccc} (2,0) & \times & (1,0) & \rightarrow & (3,0) & + & (1,1) \\ 6 & \times & 3 & = & 10 & + & 8 \end{array} \qquad (A.1b)$$

$$\begin{array}{ccccccc} (0,1) & \times & (1,0) & \rightarrow & (1,1) & + & (0,0) \\ 3 & \times & 3 & = & 8 & + & 1 \end{array} \qquad (A.1c)$$

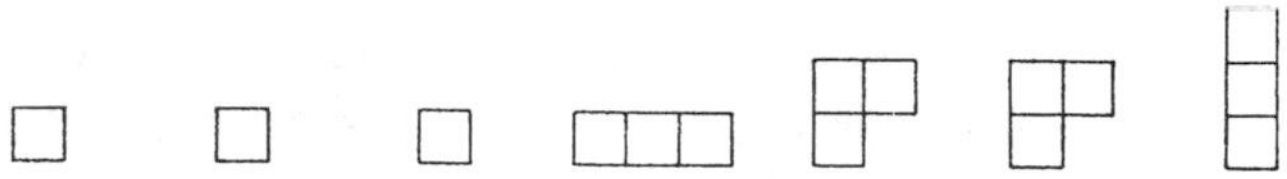

$$\begin{array}{ccccccccccccc} (1,0) & \times & (1,0) & \times & (1,0) & \rightarrow & (3,0) & + & (1,1) & + & (1,1) & + & (0,0) \\ 3 & \times & 3 & \times & 3 & = & 10 & + & 8 & + & 8 & + & 1 \end{array} \qquad (A.1d)$$

Under each multiplet is written the number of states in the multiplet, and we see that the total number of states in the product on the left-hand side of the equation checks with the total number in the sum on the right-hand side. Note that the relations (A.1) could be obtained directly from the Young diagrams without discussion of the multiplet structure.

Let us now consider all the multiplets arising in the six-sakaton system. We jump to the six-sakaton system because the particular multiplets of interest in the octet model or eightfold way arise in the six-sakaton system. The multiplets arising in the octet model all have integral values of N. Integral values of N occur in the Sakata model only when the number of sakatons is divisible by three.

The multiplets which are of particular significance are those containing at least one state of strangeness zero ($N=2$). The multiplets which contain no states of strangeness zero have $N=1$ as the largest value of N in the multiplet and are always the same as some multiplet arising in the three-sakaton system. These multiplets having no states of $N=2$ can always be broken down into a direct product of a totally antisymmetric (0,0) triplet and a multiplet arising in the three-sakaton system.

Let us first examine the states of strangeness zero ($N=2$) arising in the six-sakaton system. These are states of six nucleons which can form isospin multiplets having $T=0$, 1, 2 or 3. Since nucleon states having different values of T have different permutation symmetries, the six-nucleon states having different values of T must be in different SU_3 multiplets. We thus have four different kinds of SU_3 multiplets containing states of $N=+2$ arising in the six-sakaton system. The quantum number $\lambda=2T$ at the maximum value of N; thus these four multiplets have respectively $\lambda=0$, 2, 4 and 6. In a Young diagram describing a system of nucleons, we also have $\lambda=2T$; thus the two definitions of λ are equivalent.

Let us now examine these multiplets one at a time. The multiplet beginning with $T=3$, $\lambda=6$ has at $N=+2$, a totally symmetric state of six nucleons. It therefore has at $N=+1$, $S=-1$, a totally symmetric state of 5 nucleons and one Λ which constitutes a single isospin multiplet with $T=\frac{5}{2}$. At $N=0$ ($S=-2$) it has totally symmetric states of four nucleons and two Λ's; a single isospin multiplet with $T=2$. We can continue in this way until we reach $N=-4$ ($S=-6$) where we have a single totally symmetric state of six Λ's with $T=0$. The value of T at the minimum value of N is therefore $T=0$. Thus $\mu=0$. The multiplet has $(\lambda, \mu)=(6,0)$ and is a straightforward generalization of the (1,0), (2,0) and (3,0) multiplets which represent the totally symmetric states of the 1-, 2- and 3-sakaton systems. The multiplet diagram is an inverted triangle in which only a single value of isospin occurs at each value of N and there are no double points. Since the Young diagram for a totally symmetric state always has $\mu=0$, we see again that the definitions of μ by the Young diagram and the shape of the multiplet are equivalent. We can easily generalize this to obtain the structure of the $(n, 0)$ multiplet representing the set of totally symmetric states of an n-sakaton system.

We now can consider the multiplet beginning with $T=2$ at $N=+2$ and which has $\lambda=4$. The minimum value of N which can occur in this multiplet is $N=-3$ $(S=-5)$ corresponding to a state of five Λ's and one nucleon. The six-Λ state cannot occur here because it must be totally symmetric with regard to permutations and therefore can only be in the totally symmetric (3,0) multiplet. Since a state with only one nucleon must have $T=\frac{1}{2}$ the quantum number $\mu=2T$ at $N=N_{\text{min}}$, or $\mu=1$. Noting that the Young diagram for a six-nucleon system coupled to $T=2$ also has $\mu=1$ we again find that the definitions $(\lambda, \mu)=(4, 1)$ are equivalent. The structure of the remaining states of the multiplet can be obtained by a procedure similar to that used for obtaining the structure of the (1, 1) multiplet in the three-sakaton system. We note that the $N=+2$ state of six nucleons coupled to $T=2$ can be considered as a pair of nucleons coupled to $T=0$ and 4 nucleons coupled to $T=2$. We reach the states having $N=+1$ $(S=-1)$ by operating on these nucleon states with the operators C_+ or C_- which change a nucleon into a Λ. If we change one of the nucleons in the set of four coupled to $T=2$ into a Λ we are left with a state having $T=\frac{3}{2}$. If we change one of the nucleons in the pair coupled to $T=0$ into a Λ then its partner nucleon is now free to couple its $T=\frac{1}{2}$ to the remaining four nucleons to give $T=\frac{5}{2}$. We thus see that two values of T are possible for $N=+1$, namely, $T=\frac{5}{2}$ and $T=\frac{3}{2}$.

We now continue to the states $N=0$ $(S=-2)$. These are states of four nucleons and two Λ's. The maximum possible value of isospin is therefore $T=2$. Operating on the $S=-1$ states with the operators C_+ and C_- which change a nucleon into a Λ allows us to reach the states of $T=2$ and $T=1$ but does not allow us to reach $T=0$. By continuing this procedure we find that at $N=-1$ $(S=-3)$ we have isospin multiplets with $T=\frac{3}{2}$ and $T=\frac{1}{2}$ and that at $N=-2$ $(S=-4)$ we have isospin multiplets with $T=1$ and $T=0$. The structure of this (4, 1) multiplet thus follows the general rule that the states on the outside of the diagram are single, the next ring is double, and since this ring turns out to be a triangle, all of the states within the triangle are also double. Thus except at the maximum and minimum values of N where there is only a single isospin multiplet there are always two isospin multiplets found at each value of N.

Let us now consider the multiplet beginning with $N=+2$, $T=1$, $\lambda=2$. The six-nucleon state coupled to $T=1$ can be considered as composed of two nucleon pairs each coupled to $T=0$ and a pair coupled to $T=1$. To find the minimum value of N or strangeness occurring in this multiplet we note that the permutation symmetry limits the number of nucleons which can be changed to Λ's. Both members of an antisymmetric pair of nucleons coupled to $T=0$ cannot be changed to Λ's because a pair of Λ's must be symmetric and can only be obtained from a pair of nucleons in a symmetric state, i.e. which has $T=1$. We can therefore only make one Λ from each of the two antisymmetric nucleon pairs coupled to $T=0$ and make two more Λ's from the two nucleons coupled to $T=1$. The state of minimum N thus has four Λ's and two nucleons. It has $N=-2$, $T=1$ and therefore $\mu=2$. Since the Young diagram for a state of six nucleons having $T=2$ also has $\mu=2$, the two definitions of (λ, μ) both give (2,2) and are equivalent. The structure of this (2,2) multiplet can be obtained in the same way as the preceding ones by noting that each time we decrease N by one unit we are introducing an additional isospin of $\frac{1}{2}$ which can be coupled to the isospin of the preceding system in all possible ways, subject to the restriction that the isospin can never be greater than half the number of nucleons present at that value of strangeness. Thus at $N=+1$ we have two isospin multiplets with $T=\frac{3}{2}$ and $T=\frac{1}{2}$. At $N=0$ we have three isospin multiplets with $T=2$, 1 and 0. At $N=-1$ $(S=-3)$ there are only three nucleons present and the maximum possible value of T is $T=\frac{3}{2}$. There are therefore only two isospin multiplets having $T=\frac{3}{2}$ and $T=\frac{1}{2}$. The resulting SU_3 multiplet again satisfies the general rule: the outer ring is single, the next ring is double and the third ring consisting of a point is triple.

The final multiplet to be considered here is the one beginning with the $T=0$ at $N=+2$, and $\lambda=0$. Considering the $T=0$ six-nucleon state to be made up of three pairs individually coupled to

$T=0$ we see that we cannot change any more than three of these nucleons into Λ's and still have states of the same permutation symmetry. The minimum value of N is thus $N=-1$ ($S=-3$) and the remaining three nucleons have $T=\frac{3}{2}$ so that $\mu=3$. The Young diagram for six nucleons coupled to $T=0$ also gives $\lambda=0$, $\mu=3$, and again the two definitions are equivalent. This (0,3) multiplet is just the inverse of the triangular (3,0) multiplet found in the three-sakaton system. To verify that this is the case we must check the states arising at $N=0$ ($S=-2$) to be sure that there is only a single isospin triplet $T=1$ and no additional singlet with $T=0$. These states are obtained from the $N=-1$ ($S=-3$) state of three nucleons and three Λ's with $T=\frac{3}{2}$ by changing one of the Λ's into a nucleon. Since this means coupling an additional isospin of $\frac{1}{2}$ to $T=\frac{3}{2}$ we see that we cannot obtain a state with $T=0$. The state of $T=2$ which one might obtain in this way is excluded when one works from the top downward and notes that the $N=0$ states must also be obtainable by changing one of the nucleons in the $N=+1$, $T=\frac{1}{2}$ multiplet into a Λ.

Counting the number of states appearing in each multiplet we arrive at the following results, which we combine with previous results for the three-sakaton system.

(0,0)	1 state	$N_{max}=0$	$T=0$	at	$N=N_{max}$
(1,1)	8 states	$N_{max}=1$	$T=\frac{1}{2}$	at	$N=N_{max}$
(3,0)	10 states	$N_{max}=1$	$T=\frac{3}{2}$	at	$N=N_{max}$
(0,3)	10 states	$N_{max}=2$	$T=0$	at	$N=N_{max}$
(2,2)	27 states	$N_{max}=2$	$T=1$	at	$N=N_{max}$
(4,1)	35 states	$N_{max}=2$	$T=2$	at	$N=N_{max}$
(6,0)	28 states	$N_{max}=2$	$T=3$	at	$N=N_{max}$

The seven SU_3 multiplets listed above are all the multiplets having integral values of N with a maximum value $N_{max}=0$, 1 or 2. Using this table it is a straightforward problem to determine rules for combining those multiplets having $N_{max}=1$; the results are:

$$(1,1)+(1,1)\to(0,0)+(1,1)+(2,2)+(1,1)+(3,0)+(0,3)\,,$$
$$(1,1)+(3,0)\to(1,1)+(2,2)+(3,0)+(4,1)\,,$$
$$(3,0)+(3,0)\to(0,3)+(2,2)+(4,1)+(6,0)\,.$$

The simplest way to obtain these results is by first examining the states occurring at N_{max} and then counting states. Let us consider first the combination of (3,0) and (3,0). Since the (3,0) multiplet has ten states the combination of two of them produces a total of 100 states which we now must divide into several multiplets. The (3,0) multiplet has $N_{max}=1$ and has $T=\frac{3}{2}$ at $N=1$. The maximum value of N occurring in the combined system is then $N_{max}=2$ and the values of T occurring are 0, 1, 2 and 3. Each value of N_{max} and T determines a multiplet which must be present in the product of two (3,0) multiplets. These are respectively the (0,3), (2,2), (4,1) and (6,0). The total number of states present in these four multiplets is $10+27+35+28=100$. We have therefore accounted for all of the states obtained by combining two (3,0) multiplets and need not look for any more multiplets.

In combining two multiplets of the same kind one may be looking at states of two identical particles and therefore it is often of interest to examine the multiplets arising in the combined system from the point of view of permutation symmetry. The permutation classification is obtained immediately from the states at $N=+2$. If we are combining two states of $T=\frac{3}{2}$, the states with $T=0$ and $T=2$ are antisymmetric and the states with $T=1$ and $T=3$ are symmetric. Thus we see that the (0,3) and (4,1) multiplets contain the antisymmetric states and the (2,2) and (6,0) multiplets contain the symmetric states.

The combination of (1,1) and (3,0) can be carried through in a similar fashion. Here however at $N=2$ the values of T are obtained by combining $T=\frac{1}{2}$ from the (1,1) multiplet with $T=\frac{3}{2}$ from the (3,0) multiplet to give $T=1$ and $T=2$. We thus have a (2,2) multiplet and a (4,1) multiplet giving a total of $27+35=62$ states. The total number of states is $8\times 10=80$. Eighteen states remain to be classified. These must be in multiplets having $N_{max}\leqslant 1$ as we already have used up all the states with $N=2$. Looking at the available multiplets we see that this can be done with a (3,0) multiplet and a (1,1) multiplet giving $10+8=18$ states. One might obtain 18 states by other combinations involving a large number of (0,0) singlets. These possibilities can be eliminated by examining the total number of states having $N=1$ arising in the combination of $(1,1)$ and $(3,0)$ multiplets. In the $(1,1)$ multiplet there are two states with $N=1$ and four states with $N=0$. In the $(3,0)$ multiplet there are four states with $N=1$ and three states with $N=0$. States of $N=1$ of the combined system are obtained by all possibilities of combining the states of $N=1$ from one multiplet and $N=0$ from the other. We thus have $2\times 3+4\times 4=22$ states at $N=1$. Since the $(2,2)$ multiplet has 6 states at $N=1$ and the $(4,1)$ multiplet has 10 states at $N=1$ the combination $(1,1)+(3,0)+(2,2)+(4,1)$ has $2+4+6+10=22$ states, which checks. Other combinations in which some of these multiplets are replaced by several $(0,0)$ multiplets would not give the right number as the $(0,0)$ multiplet has no $N=1$ states. Since the $(1,1)$ and $(3,0)$ multiplets are not equivalent one cannot consider symmetric and antisymmetric combinations.

The combining of two $(1,1)$ multiplets is done in the same way. Here the values of T occurring at $N=2$ are $T=1$ and $T=0$, and give $(2,2)$ and $(0,3)$ multiplets. These account for $27+10=37$ of the total number of states, namely, $8\times 8=64$. Since the two $(1,1)$ multiplets are equivalent, we can examine permutation symmetry. The $(2,2)$ multiplet which has $T=1$ at $N=2$ contains symmetric states of the combination and the $(0,3)$ multiplet which has $T=0$ at $N=2$ contains antisymmetric states. To find the remaining multiplets in the combination we first apply similar arguments at the minimum value of N, namely $N=-2$. Here the $(1,1)$ multiplets also have states of $T=\frac{1}{2}$ at $N=-1$, giving two multiplets, one with $T=0$ and the other with $T=1$. The multiplet with $T=0$ at $N=-2$ is the $(3,0)$ multiplet and the other is the $(2,2)$ multiplet which we have already found. The $(2,2)$, $(0,3)$ and $(3,0)$ multiplets now account for $27+10+10=47$ states. There remain 17 states from the total of 64. These must be placed in multiplets which have no states with $N=+2$ or $N=-2$. This just corresponds to two $(1,1)$ multiplets and a $(0,0)$ multiplet.

Examination of the symmetries of the $N=1$ and $N=0$ states appearing in these multiplets indicate that the $(1,1)$ multiplet may contain either symmetric or antisymmetric states while the $(0,0)$ must contain a single symmetric state. It is thus convenient to group these multiplets into three symmetric multiplets, (0, 0), (1, 1) and (2, 2) and three antisymmetric multiplets, (1, 1), (3,0) and (0,3). Note however that this classification is not essential. In particular, any linear combination of the two (1, 1) multiplets is also a (1, 1) multiplet and its transformation properties under the operators of SU_3 are like any other (1, 1) multiplet. Thus two (1, 1) multiplets can be constructed by using any linear combination of states arising in the symmetric and antisymmetric multiplets. In physical problems where the two particles appearing in the two multiplets are not of the same kind there may be physical reasons why other linear combinations are more suitable for defining the (1, 1) multiplets rather than the symmetric and antisymmetric states.

These same procedures can be used to find all the SU_3 multiplets and their combination rules. The multiplets of course get larger and the calculations more unwieldy as (λ, μ) increase. In the present state of elementary-particle physics the larger multiplets are of little physical interest as there are no known particles to fit into them. Note that if any particles fitting into larger multiplets are found, a tremendous number would be necessary to fill up the multiplet. The multiplets which we have already considered include all having integral values of N and less than 50 states [except for the (0, 6) and (1, 4) multiplets which are obtained trivially by inverting the (6, 0) and (4, 1)].

The matrix elements for the operators (3.1) between states of a multiplet can be calculated in a straightforward but tedious manner from the states of the Sakata model. One can write explicit expressions for the wave functions of the states of a given number of sakatons belonging to a given SU_3 multiplet. Using these wave functions, the calculation of the matrix elements can be carried out explicitly. However, it is more convenient to obtain these matrix elements using SU_2 subgroups as is discussed in Appendix B.

APPENDIX B

CALCULATIONS OF SU_3 USING AN SU_2 SUBGROUP: U-SPIN

To make physical predictions on the basis of the SU_3 classification of elementary particles one needs coefficients for coupling the various multiplets, analogous to the Clebsch–Gordan coefficients for coupling angular momenta. The necessity for calculating new tables of coefficients can be avoided by using the SU_2 subgroups of SU_3. Since the algebra of SU_2 is the algebra of angular momentum, relations using the SU_2 subgroups involve the algebra of ordinary angular momentum and therefore the well-known and well tabulated Clebsch–Gordan coefficients. Furthermore, since most physicists are very familiar with angular momentum and have an intuitive feeling for their coupling, one can see at a glance many results for SU_2 which would either require calculation or looking at tables for SU_3.

We have seen that the algebra of SU_3 is obtained as a natural generalization of SU_2 using the Sakata model of elementary particles. The neutron, proton and lambda are all considered to be equivalent and all possible transformations of sakatons into one another are examined. Another possibility would be to consider only neutrons and lambdas and ignore the proton for the time being. Considering the algebra of all possible bilinear products of neutron and lambda operators which do not change the number of particles would lead to an SU_2 algebra directly analogous to isospin. However, instead of changing neutrons into protons and vice versa we would change neutrons into lambdas. This algebra can be characterized by an 'equivalent angular momentum' analogous to isospin, which we call

U-spin. We can use all the angular momentum results in treating U-spin, classify all states into U-spin multiplets, etc., and examine the experimental consequences of 'conservation of U-spin'.

In the framework of the Sakata model we see that U-spin conservation implies that neutrons and lambdas are equivalent. Since invariance under SU_3 implies that neutrons, lambdas and also protons are equivalent, invariance under SU_3 includes U-spin conservation as a special case. On the other hand, assuming that both U-spin and isospin are conserved implies from U-spin conservation that neutrons and lambdas are equivalent and from isospin conservation that neutrons and protons are equivalent. The two together thus imply that neutrons, protons and lambdas are all equivalent and therefore imply unitary symmetry with the full SU_3 group in the Sakata model. All the consequences of unitary symmetry in the Sakata model can therefore be obtained by examining the simultaneous consequences of isospin and U-spin conservation. Since isospin and U-spin involve only the algebra of SU_2 groups which are like angular momenta we can therefore determine all the experimental consequences of SU_3 by just coupling angular momenta of one kind or another.

We now examine the U-spin classification and the U-spin multiplets. In the Sakata model the U-spin transformations of neutrons and lambdas into one another change the strangeness of a particle but do not change the charge. U-spin multiplets therefore consist of sets of states all having the same charge and different strangeness rather than vice versa, as in the case of isospin multiplets. Since the operators (3.1) which change neutrons and lambdas into one another are the operators B_- and C_+ we can examine the operation of U-spin transformations on our multiplet diagrams. From Fig. 3.1 we see that the operators B_- and C_+ move from one state to another along a line at an angle of 120° from the horizontal isospin operators. Looking at the diagrams of Fig. 3.2 and Fig. 3.7 for the basic particles in both the Sakata and octet models of unitary symmetry we see that this direction at an angle of 120° from the isospin operators corresponds in both models to lines of constant charge.

Let us now develop the algebra of the U-spin subgroup of SU_3

without specific reference to the Sakata model so that the results can be applied to the octet model as well. We therefore define our U-spin operators directly in terms of the operators (3.1) of SU_3 and the commutation rules (3.2).

The operators B_- and C_+ are the operators that we want as the U-spin generators. Their commutator is given from eq. (3.2b)

$$[B_-, C_+] = \tfrac{1}{2}(3N - 2\tau_0) . \tag{B.1}$$

It is therefore convenient to define the U-spin operators

$$U_+ \equiv B_- , \tag{B.2a}$$

$$U_- \equiv C_+ , \tag{B.2b}$$

$$U_0 \equiv \tfrac{1}{4}(3N - 2\tau_0) . \tag{B.3}$$

Expressing the commutation rules (B.2) in terms of the U-spin operators we obtain

$$[U_+, U_-] = 2U_0 , \qquad [U_0, U_\pm] = \pm U_\pm . \tag{B.4}$$

By analogy with angular momentum we can define the total U-spin operator

$$U^2 \equiv \tfrac{1}{2}(U_+ U_- + U_- U_+) + U_0^2 , \tag{B.5a}$$

$$[U^2, U_0] = [U^2, U_+] = [U^2, U_-] = 0 . \tag{B.5b}$$

Noting from Fig. 3.1 that the operators B_- and C_+ move from one state to another along a line where $\frac{1}{2}N + \tau_0$ is constant we can define the operator

$$Q = \tfrac{1}{2}N + \tau_0 \tag{B.6a}$$

and note that

$$[U_+, Q] = [U_-, Q] = [U_0, Q] = 0 . \tag{B.6b}$$

The operator Q thus plays the same role in the U-spin representation of SU_3 as the operator N plays with regard to isospin. The operator Q commutes with all the U-spin operators and corresponds to motion on the multiplet diagram in a direction perpendicular to the direction indicated by the U-spin operators.

In the octet model where $N = Y$, i.e. the hypercharge, the operator Q defined by eq. (B.6a) is just the electric charge. In the Sakata model, $N = \frac{1}{3}B + S = Y - \frac{2}{3}B$. The operator Q defined by eq. (B.6a) is therefore not the total charge unless the baryon number is zero. However, the operator Q in the Sakata model differs from the charge only by a function of the baryon number which is the same through any multiplet. Thus in the Sakata as well as in the octet model the states in a multiplet having the same value of Q have the same charge, and U-spin transformations in both models do not change the charge. U-spin multiplets thus consist of states all having the same charge and different values of strangeness. The multiplets can be described by diagrams in which the eigenvalue of U_0 is plotted horizontally and values of Q are plotted vertically rather than plotting T_0 and N. The result is simply a rotation of the diagrams in Fig. 3.2, 3.3 and 3.6 by an angle of 120°. From the diagrams of Fig. 3.4 showing the general shape of the multiplet we see that the shape of the multiplet is left unchanged by the rotation of 120°. As a result the rotated diagrams look exactly the same as the unrotated ones except that the states in the same horizontal line now belong to the same U-spin multiplet rather than to the same isospin multiplet. Fig. B.1 illustrates this rotation for some of the common multiplets in the octet model.

For SU_3 multiplets like the (1,0), (0,1), (3,0) and (0,3) the transformation to U-spin is straightforward since there is never more than one state at a given point on the diagram. Thus the states at all points on the diagram are eigenvectors of U_0 and U^2. On the other hand, in multiplets like the (1,1) or (2,2) there are certain values of N and T_0 (or Q and U_0) where more than one state occurs. An example of this is the two states appearing at the origin of the (1,1) multiplet. We have seen that one state belongs to an isospin triplet with $T=1$ and one state corresponds to an isospin singlet with $T=0$. If we consider the (1,1) multiplet from the point of view of U-spin we find that one state must belong to a U-spin triplet with $U=1$ and one to a U-spin singlet with $U=0$. However, the two states which are eigenfunctions of U^2 are not the same two states that are eigenfunctions of T^2. We note that the operators T^2 and U^2 do not commute with one another and therefore we should not expect them to be simultaneously diagonal. For those cases where

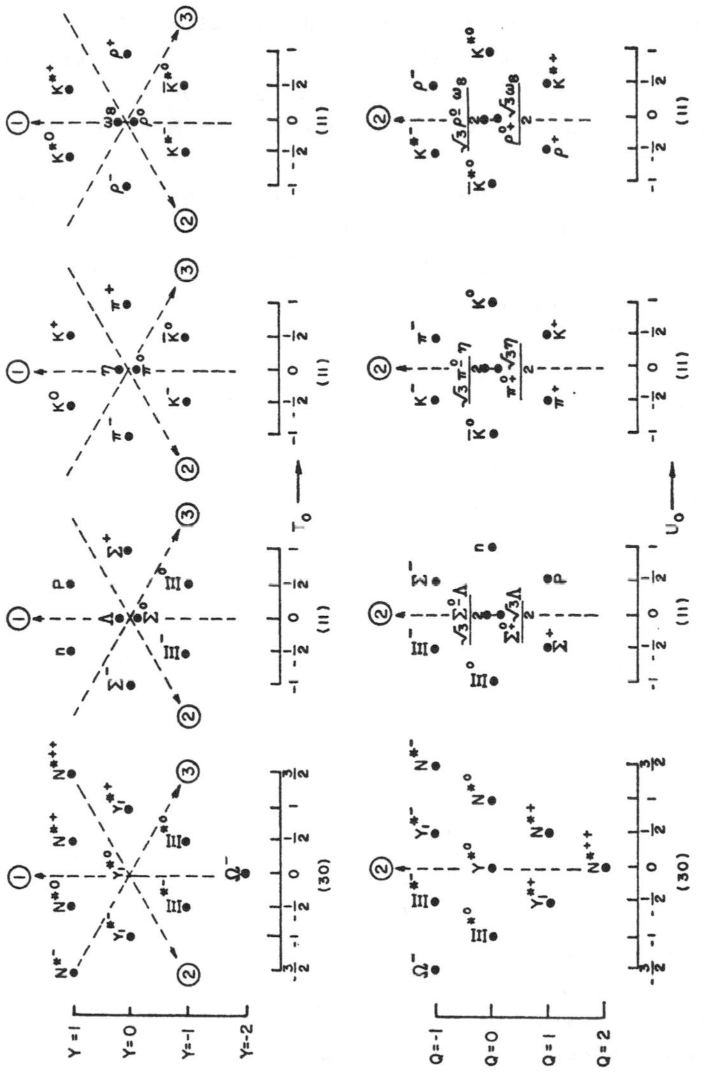

Fig. B.1

there is only one state at a given point on a diagram, i.e. only one state with a given value of N and T_0, this state must be a simultaneous eigenfunction of T^2 and U^2, since both T^2 and U^2 commute with N and T_0. If there is only one state having a given set of eigenvalues of N and T_0, the operators T^2 and U^2 cannot mix in other states. However, as soon as there is more than one state in a multiplet having a given value of N and T_0, the particular eigenfunctions which are eigenfunctions of T^2 are not eigenfunctions of U^2 and vice versa.

Fig. B.1 shows the relation between isospin and U-spin classifications of some of the elementary particles in the octet model which are classified in (1, 1) octets or the (3, 0) decuplet. Corresponding diagrams are related by a 120° rotation except for the points at the center of the octet diagram. The two states at the center of the octet which are U-spin eigenstates are linear combinations of the isospin eigenstates.

To see how the coefficients in these linear combinations are determined let us consider the baryon octet and determine the particular linear combination of the Λ and Σ^0 states which belongs in the U-spin triplet and has $U=1$. Let us call this particular linear combination

$$|U=1, U_0=0\rangle = \alpha|\Sigma^0\rangle + \beta|\Lambda\rangle \tag{B.7}$$

where α and β are constants to be determined. Since the neutron is the member of the same U-spin triplet with $U_0 = +1$ the standard lowering relation analogous to eq. (1.7) for angular momentum gives

$$U_-|n\rangle = \sqrt{2}\{\alpha|\Sigma^0\rangle + \beta|\Lambda\rangle\}\,. \tag{B.8}$$

Operating on eq. (B.8) with the isospin raising operator τ_+ then gives

$$\tau_+ U_-|n\rangle = 2\alpha|\Sigma^+\rangle\,. \tag{B.9}$$

We can now evaluate α by noting from the commutation rules (3.2b) that the operators τ_+ and U_- commute. The operation of these U-spin and isospin operators on the neutron state is easily calculated in reverse order where the intermediate state is the proton state and there are no ambiguities.

$$\tau_+ U_-|n\rangle = U_-\tau_+|n\rangle = U_-|p\rangle = |\Sigma^+\rangle\,. \tag{B.10}$$

From eq. (B.9) and eq. (B.10) we obtain

$$\alpha = \tfrac{1}{2}\,. \tag{B.11}$$

From normalization we have

$$|\beta| = \tfrac{1}{2}\sqrt{3}\,. \tag{B.12}$$

The phase of β is not uniquely determined and is fixed by convention to be positive. The U-spin singlet state at the center of the octet diagram is then determined (except for a phase factor) by requiring that it be orthogonal to the triplet state.

Note that this calculation of the values of α and β also constitutes a proof that two states are required at the center of the octet. If there were only a single state it would be a pure triplet in both isospin and U-spin and eqs. (B.9) and (B.10) would be inconsistent.

Another SU_2 subgroup in addition to isospin and U-spin can be defined by using the operators C_- and B_+. This would correspond in the Sakata model to interchanging protons and lambdas. We can again define an 'equivalent angular momentum' which we call V-spin. There the commutator of B_+ and C_- is given by eq. (3.2b),

$$[B_+, C_-] = \tfrac{1}{2}(3N+2\tau_0)\,. \tag{B.13}$$

It is therefore convenient to define the V-spin operators

$$V_+ \equiv C_-\,, \tag{B.14a}$$

$$V_- \equiv B_+\,, \tag{B.14b}$$

$$V_0 \equiv -\tfrac{1}{4}(3N+2\tau_0)\,. \tag{B.14c}$$

One could equally well have interchanged the definitions of V_+ and V_- thereby changing the sign in the definition of V_0. The particular choice made above is more symmetric with respect to the definitions of isospin and U-spin as can be seen from Fig. 3.1. A rotation by 120° carries isospin into U-spin and U-spin into V-spin.

We have now expressed all of the operators (3.1) as either isospin, U-spin or V-spin operators. It is now a comparatively simple matter to calculate the matrix elements of these operators between states of an SU_3 multiplet. They are just the matrix elements for the standard angular momentum raising or lowering operators (1.7) within isospin, U-spin or V-spin multiplets respectively. One need only to classify states in an SU_3 multiplet into U-spin and V-spin multiplets. This is trivial except for cases like the center of the (1, 1) octet where there are more than one state. The isospin, U-spin and V-spin eigenstates are generally different linear combinations of these states. However, the appropriate linear combinations can be determined in the same manner as has been done for the Λ and the Σ^0. A consistent phase convention must be chosen for all the step operator relations. This is a non-trivial problem because the conventional choice of phases in angular momentum step operator relations (1.7) cannot be used simultaneously for isospin, U-spin and V-spin without introducing inconsistencies. This point is discussed further in Appendix D.

Since all the SU_3 operators (3.1) are expressible as isospin, U-spin and V-spin operators, the requirement that any Hamiltonian be invariant under the transformations of the group SU_3 is equivalent to the requirement that it be invariant under isospin, U-spin and V-spin transformations. In fact, invariance under isospin and U-spin transformations is sufficient to require invariance under all the SU_3 transformations since V-spin is not independent of isospin and U-spin. The V-spin operators are clearly obtainable from the U-spin operators by an isospin transformation. One can therefore obtain all the results following from SU_3 invariance by requiring isospin and U-spin invariance. The latter formulation is particularly convenient since it is expressed in terms of ordinary angular momentum algebra and simply requires that both isospin and U-spin be conserved. We have chosen U-spin rather than V-spin because of its more direct physical interpretation. Particles in the same U-spin multiplet have the same value of the electric charge, whereas particles in the same V-spin multiplet have in common only a particular linear combination of electric charge and strangeness which has no direct physical significance.

Using the isospin, U-spin and V-spin operators we can now easily construct one of the Casimir operators of SU_3. If we can construct an operator which commutes with all the isospin, U-spin and V-spin operators, it commutes with all the operators of the group and satisfies the condition for being a Casimir operator. We can construct such an operator by requiring it to be a scalar under isospin rotations and to contain isospin, U-spin and V-spin in a symmetric way. We note from the commutation rules (3.2) that the operator N

is a scalar under isospin transformations and that the pairs of operators (B_+, B_-) and $(C_+, -C_-)$ transform like two-component spinors. A linear combination of the total isospin operator T^2, the operator N^2 and the scalar product of the two spinors is clearly a scalar under isospin transformations. We therefore try

$$C = \tfrac{1}{2}(\tau_+\tau_- + \tau_-\tau_+ + U_+U_- + U_-U_+ + V_+V_- + V_-V_+) + \tau_0^2 + \alpha N^2. \tag{B.15}$$

We have used $U_\pm$ and $V_\pm$ rather than $B_\pm$ and $C_\pm$ in order to show explicitly the symmetry of the operator with respect to isospin, U-spin and V-spin. The operator (B.15) is a scalar under isospin rotations and is clearly symmetric with respect to isospin, U-spin and V-spin except for the last two terms. We now wish to choose the coefficient α to complete this symmetry. This is easily done by noting from the definitions (B.3) and (B.14c) of U_0 and V_0 that

$$\tau_0^2 + U_0^2 + V_0^2 = \tfrac{9}{8}N^2 + \tfrac{3}{2}\tau_0^2\,. \tag{B.16}$$

Setting $\alpha = \frac{3}{4}$ and using eq. (B.16) gives

$$\begin{aligned} C &= \tfrac{1}{2}(\tau_+\tau_- + \tau_-\tau_+ + U_+U_- + U_-U_+ + V_+V_- + V_-V_+) + \\ &\qquad + \tfrac{2}{3}(\tau_0^2 + U_0^2 + V_0^2) \\ &= T^2 + U^2 + V^2 - \tfrac{1}{3}(\tau_0^2 + U_0^2 + V_0^2)\,. \end{aligned} \tag{B.17}$$

From the form (B.17) we see that the operator C is completely symmetric in isospin, U-spin and V-spin, while from the form (B.15) we see that it is a scalar under isospin transformations. It therefore commutes with all the isospin operators and by symmetry with all the U-spin and V-spin operators, and is thus a Casimir operator for the group SU_3. Comparing this result with (4.12) we see that we have found the same Casimir operator, but have chosen a different normalization.

From the expression (B.17) we can easily calculate the eigenvalue of the operator C for any multiplet (λ, μ). Since all members of an SU_3 multiplet are eigenfunctions of C with the same eigenvalue, we pick a convenient member, namely the state at the right-hand end of the top row in the multiplet diagram. This state is a simultaneous eigenfunction of isospin, U-spin and V-spin with $T = \frac{1}{2}\lambda$, $U = \frac{1}{2}\mu$ and $V = \frac{1}{2}(\lambda + \mu)$. This state is also at the end of all these spin multiplets and has $|T_0|=T$, $|U_0|=U$ and $|V_0|=V$. We thus obtain

$$\begin{aligned} C(\lambda, \mu) &= T(T+1)+U(U+1)+V(V+1)-\tfrac{1}{3}(T^2+U^2+V^2) \\ &= \tfrac{2}{3}\{\tfrac{1}{4}\lambda^2+\tfrac{1}{4}\mu^2+\tfrac{1}{4}(\lambda+\mu)^2\}+\{\tfrac{1}{2}\lambda+\tfrac{1}{2}\mu+\tfrac{1}{2}(\lambda+\mu)\} \\ &= \tfrac{1}{3}\{(\lambda-\mu)^2\}+\lambda+\mu+\lambda\mu\,. \end{aligned} \tag{B.18}$$

This result is a complicated second-degree polynomial in λ and μ. It is therefore more convenient to use λ and μ as labels to specify an SU_3 multiplet since these take on simple integral values rather than the complicated set of eigenvalues of the Casimir operator (B.18). Note that the eigenvalues (B.18) have the property of being either an integer or a third integer depending upon whether or not $\frac{1}{3}(\lambda-\mu)$ is an integer. This supports our previous classification of multiplets according to the occurrence of integral or third-integral eigenvalues for N.

APPENDIX C

EXPERIMENTAL PREDICTIONS FROM THE OCTET MODEL OF UNITARY SYMMETRY

We shall now examine in detail with examples the experimental predictions outlined in § 3.6 which can be made on the basis of the octet model of SU_3. Predictions of type A and B follow from the assumption that the interactions between strongly interacting particles are invariant to a good approximation under the transformations of the group SU_3 with the octet classification for the particles. Predictions of type C follow from the assumption that the symmetry-breaking interactions transform in a particular way under SU_3.

A. Predictions following from the multiplet structure

1. *Classification.* All particles and resonances must fit into SU_3 multiplets. Thus when only the π and K pseudoscalar mesons were known, SU_3 predicted the existence of the η to complete the octet. The eight baryons (N, Σ, Λ, Ξ) just fit into an octet and the nine vector meson resonances just fit into an octet and a singlet. The physical φ- and ω-mesons both have ($T=0$, $Y=0$) and are mixtures of the unitary singlet and octet. This mixing is apparently due to the symmetry-breaking part of the strong interaction. The linear combinations of the physical φ- and ω-mesons which belong to the singlet and octet are denoted by ω_1 and ω_8. A mixing parameter λ is defined by the relations:

$$|\omega_1\rangle = \cos\lambda|\omega\rangle + \sin\lambda|\varphi\rangle\,, \tag{C.1a}$$

$$|\omega_8\rangle = -\sin\lambda|\omega\rangle + \cos\lambda|\varphi\rangle\,. \tag{C.1b}$$

The vector meson octet thus includes (K*, ϱ, ω_8) and the singlet just the ω_1. The cause and nature of the ω–φ mixing is not discussed further in this treatment.

The $p_{\frac{3}{2}}$ meson–baryon resonances $N^*(T=\frac{3}{2}, Y=1)$, $Y^*(T=1, Y=0)$ and $\Xi^*(T=\frac{1}{2}, Y=-1)$ suggested a (3,0) decuplet with the prediction of the existence of the $\Omega^-(T=0, Y=-2)$. The first three resonances could also have been fitted into a 27-dimensional (2,2) multiplet with other predictions; namely two $T=1$ triplets at $Y=\pm 2$, an additional $T=\frac{1}{2}$ doublet at $Y=1$ and a $T=\frac{3}{2}$ quartet at $Y=-1$, and a $T=0$ singlet and a $T=2$ quintet at $Y=0$. Before the discovery of the Ω^-, the decuplet (3,0) classification was already indicated by the experimental absence of any K–nucleon resonance which would be necessary for the $Y=+2$ state in the 27.

2. *Couplings.* All resonances must be classified in multiplets which can arise in the coupling of the constituents. Thus all meson–baryon resonances must fit into the multiplets arising in the coupling of two (1,1) octets, namely the (0,0) singlet, the (1,1) octet, the (3,0) and (0,3) decuplets and the 27-dimensional (2,2). From this it follows immediately that the $p_{\frac{3}{2}}$ meson–baryon resonances discussed above can only be placed into (3,0) or (2,2) multiplets, as these are the only multiplets containing a $T=\frac{3}{2}$ state which can be obtained by the coupling of two octets.

Any new baryon–meson or meson–meson resonances must also be classified in the same way. Thus any $(T=1, Y=0)$ resonance must belong either to an octet, a decuplet or a 27; a $(T=\frac{1}{2}, Y=\pm 1)$ to an octet, decuplet or 27, and a $(T=\frac{3}{2}, Y=\pm 1)$ to a decuplet or 27. A $Y=\pm 2$ resonance must belong to a decuplet if it has $T=0$, or to a 27 if it has $T=1$.

B. Relations between matrix elements

These predictions all follow from the Wigner–Eckart theorem, using the generalized Clebsch–Gordan coefficients for the group SU_3. We shall use the SU_2 subgroups of isospin and U-spin in order to obtain these results with ordinary angular momentum coupling algebra, i.e. we shall require conservation of isospin and U-spin.

1. *Decay widths.* The decay of any resonance into two particles is

related to corresponding decays of resonances in the same SU_3 multiplet. The relations involve a single coupling, that of the two multiplets of the final state to make a multiplet like that of the initial state. In general, all decays are proportional to one another, with coefficients which are SU_3 coupling coefficients. There are exceptional cases in which there is more than one possible coupling. The only one of practical interest for elementary particles is the coupling of two octets to make an octet. This arises in the decay of meson–baryon resonances belonging to an octet. In such cases there are two independent 'reduced matrix elements' instead of only one. In meson–meson resonances this does not occur, because of the requirement of charge conjugation invariance, as has been discussed in section 3.4.

We first consider the decay of a baryon resonance in the (3,0) decuplet (N^*, Y^*, Ξ^*, Ω^-) into a baryon and a pseudoscalar meson, the latter being both in (1,1) octets. If the interaction giving rise to this decay is invariant under SU_3 all the matrix elements are expressible in terms of a single parameter. The relations between the different decays would be given by generalized Clebsch–Gordan coefficients involving the coupling of two octets to make a decuplet. We can however obtain all these results from U-spin and isospin conservation and ordinary angular momentum Clebsch–Gordan coefficients. We shall evaluate matrix elements for all possible decays which conserve isospin and strangeness, without regard for the physical masses of the particles. We therefore also consider decays such as $\langle N^*|\Sigma K\rangle$ which are impossible in practice because of energy conservation and mass differences. (Such matrix elements may still be of physical interest as describing virtual transitions or vertices in Feynman diagrams with some particles off the mass shell.) Since all the masses of the particles in a given multiplet would be the same under pure SU_3 invariance, the effects of mass differences on the decays must be introduced separately on the basis of dynamical considerations extraneous to SU_3, such as differences in phase space and barrier penetration factors.

We consider first the decay of the negatively charged baryon resonances (N^{*-}, Y^{*-}, Ξ^{*-} and Ω^-) into a negative baryon

(Σ^-, Ξ^-) and a neutral meson (K^0, π^0, η, $\overline{K}^0$). The negative excited baryon resonances in the decuplet form a U-spin quartet with $U=\frac{3}{2}$. The negative baryons form a U-spin doublet with $U=\frac{1}{2}$. The neutral mesons form a triplet and a singlet with $U=1$ and $U=0$ respectively. We see immediately that the U-spin singlet meson cannot contribute to this decay since one cannot couple a U-spin singlet to a U-spin doublet to get $U=\frac{3}{2}$. Thus only the particular linear combination of π^0 and η which has $U=1$ contributes to this decay. We can therefore write the transition matrix element for all of these decays of a negative baryon resonance to a negative baryon and a neutral pseudoscalar meson in terms of a single amplitude A with coefficients involving the coupling of a spin $\frac{1}{2}$ and a spin 1 to obtain a spin $\frac{3}{2}$. The results are:

$$\langle N^{*-}|\Sigma^- K^0\rangle = (\tfrac{1}{2}1 \ \tfrac{1}{2} \ 1|\tfrac{3}{2} \ \tfrac{3}{2})A\,, \tag{C.2a}$$

$$\langle Y^{*-}|\Xi^- K^0\rangle = (\tfrac{1}{2}1 -\tfrac{1}{2} \ 1|\tfrac{3}{2} \ \tfrac{1}{2})A\,, \tag{C.2b}$$

$$\langle Y^{*-}|\Sigma^- \pi^0\rangle = \tfrac{1}{2}(\tfrac{1}{2}1 \ \tfrac{1}{2} \ 0|\tfrac{3}{2} \ \tfrac{1}{2})A\,, \tag{C.2c}$$

$$\langle Y^{*-}|\Sigma^- \eta\rangle = \tfrac{1}{2}\sqrt{3}(\tfrac{1}{2}1 \ \tfrac{1}{2} \ 0|\tfrac{3}{2} \ \tfrac{1}{2})A\,, \tag{C.2d}$$

$$\langle \Xi^{*-}|\Sigma^- \overline{K}^0\rangle = (\tfrac{1}{2}1 \ \tfrac{1}{2} -1|\tfrac{3}{2} -\tfrac{1}{2})A\,, \tag{C.2e}$$

$$\langle \Xi^{*-}|\Xi^- \pi^0\rangle = \tfrac{1}{2}(\tfrac{1}{2}1 -\tfrac{1}{2} \ 0|\tfrac{3}{2} -\tfrac{1}{2})A\,, \tag{C.2f}$$

$$\langle \Xi^{*-}|\Xi^- \eta\rangle = \tfrac{1}{2}\sqrt{3}(\tfrac{1}{2}1 -\tfrac{1}{2} \ 0|\tfrac{3}{2} -\tfrac{1}{2})A\,, \tag{C.2g}$$

$$\langle \Omega^-|\Xi^- \overline{K}^0\rangle = (\tfrac{1}{2}1 -\tfrac{1}{2} -1|\tfrac{3}{2} -\tfrac{3}{2})A\,. \tag{C.2h}$$

A similar analysis can be made for the decay of the negative baryon resonances into a neutral baryon and a negative meson. Here again there is a single amplitude which we can call B and the Clebsch–Gordan coefficients again involve the coupling of a spin $\frac{1}{2}$ and a spin 1 to get a total spin of $\frac{3}{2}$. However, in this case it is the neutral baryon which has the U-spin 1 and the negative meson which has the U-spin $\frac{1}{2}$.

$$\langle N^{*-}|\pi^- n\rangle = (\tfrac{1}{2}1 \ \tfrac{1}{2} \ 1|\tfrac{3}{2} \ \tfrac{3}{2})B\,, \tag{C.3a}$$

$$\langle Y^{*-}|K^- n\rangle = (\tfrac{1}{2}1 -\tfrac{1}{2} \ 1|\tfrac{3}{2} \ \tfrac{1}{2})B\,, \tag{C.3b}$$

$$\langle Y^{*-}|\pi^- \Sigma^0\rangle = \tfrac{1}{2}(\tfrac{1}{2}1 \ \tfrac{1}{2} \ 0|\tfrac{3}{2} \ \tfrac{1}{2})B\,, \tag{C.3c}$$

$$\langle Y^{*-}|\pi^- \Lambda\rangle = \tfrac{1}{2}\sqrt{3}(\tfrac{1}{2}1 \ \tfrac{1}{2} \ 0|\tfrac{3}{2} \ \tfrac{1}{2})B\,, \tag{C.3d}$$

$$\langle \Xi^{*-} | \pi^- \, \Xi^0 \rangle = (\tfrac{1}{2}\,1 \quad \tfrac{1}{2}\,-1|\tfrac{3}{2}\,-\tfrac{1}{2})B\,, \tag{C.3e}$$

$$\langle \Xi^{*-} | \mathrm{K}^- \, \Sigma^0 \rangle = \tfrac{1}{2}(\tfrac{1}{2}\,1 \, -\tfrac{1}{2} \quad 0|\tfrac{3}{2}\,-\tfrac{1}{2})B\,, \tag{C.3f}$$

$$\langle \Xi^{*-} | \mathrm{K}^- \, \Lambda \rangle = \tfrac{1}{2}\sqrt{3}(\tfrac{1}{2}\,1 \, -\tfrac{1}{2} \quad 0|\tfrac{3}{2}\,-\tfrac{1}{2})B\,, \tag{C.3g}$$

$$\langle \Omega^- | \mathrm{K}^- \, \Xi^0 \rangle = (\tfrac{1}{2}\,1 \, -\tfrac{1}{2}\,-1|\tfrac{3}{2}\,-\tfrac{3}{2})B\,. \tag{C.3h}$$

The amplitudes A and B are not independent but are related by isospin considerations. The simplest way to obtain this relation is from the Ω^- which has $T=0$ and goes to the opposite members of two isospin doublets, namely $|\mathrm{K}^- \Xi^0\rangle$ and $|\overline{\mathrm{K}}{}^0 \Xi^-\rangle$. Since the linear combination $|\mathrm{K}^- \Xi^0\rangle - |\overline{\mathrm{K}}{}^0 \Xi^-\rangle$ has $T=0$, we obtain $A = -B$. The same result is obtained from the $\langle \mathrm{Y}^{*-}|\Sigma^- \pi^0\rangle$ and $\langle \mathrm{Y}^{*-}|\Sigma^0 \pi^-\rangle$ amplitudes which go through a single isospin channel with $T=1$.

The decays for the other charge states of the baryon resonances are simply related to the above decays by isospin considerations.

We next consider the decay of a vector meson in the (1,1) octet into two pseudoscalar mesons in (1,1) octets. If the interaction giving rise to this decay is invariant under SU_3 all the matrix elements are not necessarily expressible in terms of a single parameter because there are two ways to couple two octets to make an octet. However, the requirement of invariance either under space inversion or charge conjugation is sufficient to eliminate the symmetric coupling. The two pseudoscalar mesons must be in a state of odd parity if parity is conserved in the decay of the vector meson. An odd parity state of two spinless particles is also odd under permutation of the two particles. The generalized Pauli principle for bosons therefore requires them to be in an SU_3 multiplet which is antisymmetric under the permutation of the two octets. The symmetric coupling is therefore excluded. The same result follows from charge conjugation invariance by noting that the ϱ^0 and ω_8 are odd under charge conjugation and decay into two pseudoscalar mesons which are charge conjugates of one another. Charge conjugation in the final state is thus equivalent to permutation of the two particles and the state must be odd under this operation.

Since only the antisymmetric octet coupling is allowed, all the decay matrix elements are again expressible in terms of a single parameter. The relations between the different decays are again obtainable from U-spin and isospin conservation using ordinary angular momentum Clebsch–Gordan coefficients with the additional requirements of parity conservation.

Consider first the decay of a neutral vector meson into two charged pseudoscalar mesons. The K^{*0}, the $\overline{K}^{*0}$, and the linear combination $\{\frac{1}{2}\varrho^0+\frac{1}{2}\sqrt{3}\omega_8\}$ form a U-spin triplet with $U=1$. The charged π- and K-mesons form two U-spin doublets. We thus obtain

$$\langle\{\tfrac{1}{2}\varrho^0+\tfrac{1}{2}\sqrt{3}\omega_8\}|K^+K^-\rangle:\langle\{\tfrac{1}{2}\varrho^0+\tfrac{1}{2}\sqrt{3}\omega_8\}|\pi^+\pi^-\rangle:\langle K^{*0}|K^+\pi^-\rangle$$
$$=(\tfrac{1}{2}\tfrac{1}{2}\tfrac{1}{2}-\tfrac{1}{2}|1\,0):(\tfrac{1}{2}\tfrac{1}{2}-\tfrac{1}{2}\tfrac{1}{2}|1\,0):(\tfrac{1}{2}\tfrac{1}{2}\tfrac{1}{2}\tfrac{1}{2}|1\,1)\,. \qquad \text{(C.4)}$$

We now note that since the two pions are spinless bosons in the same isospin multiplet, they can be in an antisymmetric spatial state only if they are antisymmetric in isospin, i.e., only if they have $T=1$. To make a vector meson with odd parity, the two pions must be in an antisymmetric state. They therefore have $T=1$. A $T=0$ vector meson therefore cannot decay into two pions. Thus

$$\langle\omega_8|\pi^+\pi^-\rangle=0\,. \qquad \text{(C.5)}$$

This result can also be obtained directly from conservation of *G*-parity. Discarding the term describing the decay of the ω_8 into two pions leaves

$$\langle K^{*0}|K^+\pi^-\rangle=\tfrac{1}{2}\sqrt{2}\langle\varrho^0|\pi^+\pi^-\rangle\,, \qquad \text{(C.6a)}$$

$$\langle\tfrac{1}{2}\varrho^0+\tfrac{1}{2}\sqrt{3}\,\omega_8|K^+K^-\rangle=-\tfrac{1}{2}\langle\varrho^0|\pi^+\pi^-\rangle\,. \qquad \text{(C.6b)}$$

The first of these relations allows us to predict the ratio of the width of the K^{*0} to the width of the ϱ. To do this we need to introduce the neutral decay mode of the K^{*0} which is related to the charged mode by isospin coupling. This relation is

$$\frac{\langle K^{*0}|K^0\pi^0\rangle}{\langle K^{*0}|K^+\pi^-\rangle}=\frac{(\tfrac{1}{2}1-\tfrac{1}{2}\quad 0|\tfrac{1}{2}-\tfrac{1}{2})}{(\tfrac{1}{2}1\quad \tfrac{1}{2}-1|\tfrac{1}{2}-\tfrac{1}{2})}=-\frac{1}{\sqrt{2}}\,. \qquad \text{(C.7a)}$$

Thus

$$\langle K^{*0}|K^0\pi^0\rangle=-\tfrac{1}{2}\langle\varrho^0|\pi^+\pi^-\rangle \qquad \text{(C.7b)}$$

and

$$\frac{|\langle K^{*0}|K^+\pi^-\rangle|^2+|\langle K^{*0}|K^0\pi^0\rangle|^2}{|\langle\varrho^0|\pi^+\pi^-\rangle|^2}=\tfrac{1}{2}+\tfrac{1}{4}=\tfrac{3}{4}\,. \qquad \text{(C.7c)}$$

To obtain the ratio of the K*- and ϱ-widths, the factor $\frac{3}{4}$ must be multiplied by phase space factors to include the effect of the different momenta of the two final states. The result is in reasonable agreement with the experimental value of this ratio, which is about $\frac{1}{2}$.

Let us now consider the decay of neutral vector mesons into K^0 and $\overline{K}^0$. The neutral K-mesons are members of the same $U=1$ multiplet and must again be in the space-antisymmetric p-state. By an argument directly analogous to the isospin argument employed above for the two-pion system, we see that the $K^0\overline{K}^0$ system in an antisymmetric p-state must also be antisymmetric in U-spin and must have $U=1$. Thus only the $U=1$ linear combination $\frac{1}{2}\varrho^0+\frac{1}{2}\sqrt{3}\,\omega_8$ is coupled to the $K^0\overline{K}^0$ system and we obtain

$$\langle\varrho^0|K^0\overline{K}^0\rangle=\tfrac{1}{3}\sqrt{3}\langle\omega_8|K^0\overline{K}^0\rangle\,. \tag{C.8}$$

This result can also be expressed in terms of the charged-K mode as

$$\langle\varrho^0|K^+K^-\rangle=-\tfrac{1}{3}\sqrt{3}\langle\omega_8|K^+K^-\rangle\,, \tag{C.9}$$

where the negative sign comes from isospin coupling of $\frac{1}{2}+\frac{1}{2}$ to 1 on the left-hand side and to zero on the right. Then by substituting eq. (C.9) into eq. (C.6b) we can relate the φ-decay to the ϱ-decay through the expression

$$\langle\omega_8|K^+K^-\rangle=-\tfrac{1}{2}\sqrt{3}\langle\varrho^0|\pi^+\pi^-\rangle\,. \tag{C.10}$$

These results expressed in terms of the ω_8 are easily expressed in terms of the physical φ and ω vector mesons by noting that only the ω_8 linear combination of the physical φ and ω can contribute to this decay. The unitary singlet has $U=0$ and is not coupled to the $K^0\overline{K}^0$ system which has $U=1$. Thus

$$\langle\varphi|K^+K^-\rangle=-\tfrac{1}{2}\sqrt{3}\cos\lambda\,\langle\varrho^0|\pi^+\pi^-\rangle\,, \tag{C.11a}$$

$$\langle\varrho^0|K^+K^-\rangle:\langle\varphi|K^+K^-\rangle:\langle\omega|K^+K^-\rangle=$$
$$=1:-\sqrt{3}\cos\lambda:\sqrt{3}\sin\lambda\,. \tag{C.11b}$$

Eq. (C.11a) relates the width of the φ to the width of the ϱ. Eq. (C.11b) can be interpreted as giving the ratio of the production matrix elements for ϱ^0, φ and ω production by a $K\overline{K}$-vertex. For example, in K^-p reactions which go via one-K exchange, the ratio is

$$\langle K^- p|\varrho^0 \Lambda\rangle : \langle K^- p|\omega\Lambda\rangle : \langle K^- p|\varphi\Lambda\rangle = 1 : \sqrt{3}\sin\lambda : -\sqrt{3}\cos\lambda . \tag{C.12}$$

2. *Relations between cross sections.* We first consider the production of a baryon resonance in the (3,0) decuplet together with a pseudoscalar meson in a meson–baryon reaction

$$\begin{array}{lcccc} & \mathrm{M} \quad + & \mathrm{B} \rightarrow & \mathrm{M} \quad + & \mathrm{B}^* \\ (\lambda,\mu) & (1,1) & (1,1) & (1,1) & (3,0) \\ \text{Combined } (\lambda,\mu) & \multicolumn{2}{c}{\underbrace{(0,0),\ (1,1),\ (1,1)\ (2,2),\ (3,0),\ (0,3)}} & \multicolumn{2}{c}{\underbrace{(1,1),\ (2,2),\ (3,0),\ (4,1)}} \end{array} \tag{C.13}$$

The meson and baryon in the initial state are both in (1,1) octets and these can be coupled to any of the six possible multiplets arising in the coupling of two octets. The meson in the final state is also in a (1,1) octet and can be coupled with the baryon resonance to any of the four multiplets arising in the coupling of a (1,1) octet with a (3,0) decuplet. If the interaction giving rise to this reaction is invariant under SU_3 each multiplet which is common to both sides of the reaction can define a channel through which the reaction can occur. We see that the (1,1), (2,2) and the (3,0) multiplets are all possible. Since there are two independent (1,1) octets in the initial state there are two independent octet channels. The total number of independent channels is therefore four and all reactions of this type are therefore expressible in terms of four independent complex amplitudes. These are obtainable directly using the appropriate generalized Clebsch–Gordan coefficients for these couplings.

All the relations obtainable with SU_3 couplings are obtainable using U-spin and isospin conservation as in the previous examples. In this case there is an additional advantage to the use of U-spin. Relations between different reaction cross sections involving four independent complex amplitudes (i.e. seven independent real parameters) are not generally very useful. One would like to find several reactions expressible in terms of only a single parameter as in the decay widths above, or in terms of a comparatively small number of parameters. Using the SU_3 couplings there is no easy way to find such simple relations. One simply has to evaluate the transition amplitudes for all reactions in terms of the four channel amplitudes and then look for those which happen to be proportional to the same linear combinations of channel amplitudes. Using U-spin it is possible to find almost immediately certain reactions which are simply related. One can also restrict attention to those reactions which are of practical interest; i.e. which require nucleon rather than hyperon targets for the initial state.

Consider the four reactions in which a negative meson incident on a proton produces a positive meson and a negatively charged baryon resonance.

$$\begin{array}{lcccc} & \pi^- + p & \rightarrow & K^+ + Y_1^{*-}, \\ & \pi^- + p & \rightarrow & \pi^+ + N^{*-}, \\ & K^- + p & \rightarrow & K^+ + \Xi^{*-}, \\ & K^- + p & \rightarrow & \pi^+ + Y_1^{*-}. \\ U: & \underbrace{\tfrac{1}{2} \quad \tfrac{1}{2}} & & \underbrace{\tfrac{1}{2} \quad \tfrac{3}{2}}, \\ U_{\text{total}}: & 0 \text{ or } 1 & & 1 \text{ or } 2. \end{array}$$

The π^- and K^- belong to the same U-spin doublet with $U=\frac{1}{2}$. The proton is a member of a U-spin doublet with $U=\frac{1}{2}$. There are thus two possible U-spin states for the left-hand side of these reactions, namely $U=0$ and $U=1$. The K^+ and π^+ are members of the same U-spin doublet with $U=\frac{1}{2}$. The Y^{*-}, N^{*-} and Ξ^{*-} are all members of a U-spin quartet with $U=\frac{3}{2}$. The possible U-spin states obtained from coupling $U=\frac{1}{2}$ with $U=\frac{3}{2}$ are $U=1$ and $U=2$. If U-spin is conserved in these reactions, there is thus only one possible U-spin channel that is common to both sides of the reactions, namely $U=1$. The amplitudes for these four reactions are therefore all expressed in terms of a single parameter, the amplitude for the $U=1$ channel. The coefficients are products of two Clebsch–Gordan coefficients, one for each side of a reaction. The coefficient for the left-hand side describes the coupling of two spins of $\frac{1}{2}$ to a total spin 1; the one for the right-hand side describes the coupling of a spin of $\frac{1}{2}$ and a spin of $\frac{3}{2}$ to a total spin 1. These amplitudes are

$$\begin{aligned}
\langle\pi^- \, \mathrm{p}|\mathrm{K}^+ \mathrm{Y}_1^{*-}\rangle &= (\tfrac{1}{2}\tfrac{1}{2} \quad \tfrac{1}{2}\tfrac{1}{2}|1\,1)(\tfrac{1}{2}\tfrac{3}{2} \quad \tfrac{1}{2} \quad \tfrac{1}{2}|1\,1)a_1 = -\tfrac{1}{2}a_1\,, \\
\langle\pi^- \, \mathrm{p}|\pi^+ \, \mathrm{N}^{*-}\rangle &= (\tfrac{1}{2}\tfrac{1}{2} \quad \tfrac{1}{2}\tfrac{1}{2}|1\,1)(\tfrac{1}{2}\tfrac{3}{2} -\tfrac{1}{2} \quad \tfrac{3}{2}|1\,1)a_1 = \tfrac{1}{2}\sqrt{3}a_1\,, \\
\langle\mathrm{K}^- \mathrm{p}|\mathrm{K}^+ \, \Xi^{*-}\rangle &= (\tfrac{1}{2}\tfrac{1}{2} -\tfrac{1}{2}\tfrac{1}{2}|1\,0)(\tfrac{1}{2}\tfrac{3}{2} \quad \tfrac{1}{2} -\tfrac{1}{2}|1\,0)a_1 = -\tfrac{1}{2}a_1\,, \\
\langle\mathrm{K}^- \mathrm{p}|\pi^+ \, \mathrm{Y}^{*-}\rangle &= (\tfrac{1}{2}\tfrac{1}{2} -\tfrac{1}{2}\tfrac{1}{2}|1\,0)(\tfrac{1}{2}\tfrac{3}{2} -\tfrac{1}{2} \quad \tfrac{1}{2}|1\,0)a_1 = \quad \tfrac{1}{2}a_1\,.
\end{aligned} \tag{C.14}$$

We next consider the production in proton–antiproton annihilation of a baryon resonance in the (3,0) decuplet together with its corresponding antiparticle in the (0,3) decuplet. Here again, examination of the SU_3 couplings shows four independent channels. Using U-spin, one finds sets of relations involving only two channels since the proton and antiproton are both in U-spin doublets having $U=\frac{1}{2}$ and can couple only to a total U of 1 or 0.

Consider the production of negatively charged baryon resonances and their corresponding positively charged antiparticles.

$$\begin{aligned}
&\bar{\mathrm{p}}+\mathrm{p} \to \mathrm{N}^{*-} + \overline{\mathrm{N}^{*-}}\,, \\
&\bar{\mathrm{p}}+\mathrm{p} \to \mathrm{Y}^{*-} + \overline{\mathrm{Y}^{*-}}\,, \\
&\bar{\mathrm{p}}+\mathrm{p} \to \Xi^{*-} + \overline{\Xi^{*-}}\,, \\
&\bar{\mathrm{p}}+\mathrm{p} \to \Omega^{-} + \overline{\Omega^{-}}\,. \\
U\!:\quad &\underbrace{\tfrac{1}{2} \;\; \tfrac{1}{2}} \quad \underbrace{\tfrac{3}{2} \qquad\quad \tfrac{3}{2}}\,, \\
U_{\text{total}}\!:\quad &\;0 \text{ or } 1 \quad\; 0,1,2 \text{ or } 3\,.
\end{aligned}$$

The proton and antiproton are both in U-spin doublets having $U=\frac{1}{2}$. The reactions above therefore go through two U-spin channels ($U=0$ and $U=1$) and are expressed in terms of two independent complex amplitudes. Letting a_0 and a_1 be the amplitudes for the $U=0$ and $U=1$ channels, we can write

$$\begin{aligned}
(\bar{\mathrm{p}}\,\mathrm{p}|\mathrm{N}^{*-}\,\overline{\mathrm{N}^{*-}}) &= -\sqrt{\tfrac{1}{2}}(\tfrac{3}{2}\tfrac{3}{2} \quad \tfrac{3}{2} -\tfrac{3}{2}|0\,0)a_0 + \sqrt{\tfrac{1}{2}}(\tfrac{3}{2}\tfrac{3}{2} \quad \tfrac{3}{2} -\tfrac{3}{2}|1\,0)a_1\,, \\
(\bar{\mathrm{p}}\,\mathrm{p}|\mathrm{Y}^{*-}\,\overline{\mathrm{Y}^{*-}}) &= -\sqrt{\tfrac{1}{2}}(\tfrac{3}{2}\tfrac{3}{2} \quad \tfrac{1}{2} -\tfrac{1}{2}|0\,0)a_0 + \sqrt{\tfrac{1}{2}}(\tfrac{3}{2}\tfrac{3}{2} \quad \tfrac{1}{2} -\tfrac{1}{2}|1\,0)a_1\,, \\
(\bar{\mathrm{p}}\,\mathrm{p}|\Xi^{*-}\,\overline{\Xi^{*-}}) &= -\sqrt{\tfrac{1}{2}}(\tfrac{3}{2}\tfrac{3}{2} -\tfrac{1}{2} \quad \tfrac{1}{2}|0\,0)a_0 + \sqrt{\tfrac{1}{2}}(\tfrac{3}{2}\tfrac{3}{2} -\tfrac{1}{2} \quad \tfrac{1}{2}|1\,0)a_1\,, \\
(\bar{\mathrm{p}}\,\mathrm{p}|\Omega^{-}\,\overline{\Omega^{-}}) &= -\sqrt{\tfrac{1}{2}}(\tfrac{3}{2}\tfrac{3}{2} -\tfrac{3}{2} \quad \tfrac{3}{2}|0\,0)a_0 + \sqrt{\tfrac{1}{2}}(\tfrac{3}{2}\tfrac{3}{2} -\tfrac{3}{2} \quad \tfrac{3}{2}|1\,0)a_1\,,
\end{aligned} \tag{C.15}$$

where the factors $\pm\sqrt{\frac{1}{2}}$ come from coupling $(\bar{\mathrm{p}}\,\mathrm{p})$ to $U=1$ and $U=0$.

When the numerical values for the Clebsch–Gordan coefficients are inserted, these become

$$
\begin{aligned}
\sqrt{2}(\bar{p}p|N^{*-}\overline{N^{*-}}) &= -\tfrac{1}{2}a_0+\tfrac{3}{2\sqrt{5}}a_1\,, \\
\sqrt{2}(\bar{p}p|Y^{*-}\overline{Y^{*-}}) &= +\tfrac{1}{2}a_0-\tfrac{1}{2\sqrt{5}}a_1\,, \\
\sqrt{2}(\bar{p}p|\Xi^{*-}\overline{\Xi^{*-}}) &= -\tfrac{1}{2}a_0-\tfrac{1}{2\sqrt{5}}a_1\,, \\
\sqrt{2}(\bar{p}p|\Omega^{-}\overline{\Omega^{-}}) &= +\tfrac{1}{2}a_0+\tfrac{3}{2\sqrt{5}}a_1\,.
\end{aligned} \tag{C.16}
$$

These four relations involve three parameters: the magnitudes of the amplitudes a_0 and a_1 and the relative phase. Eliminating these three parameters between the four equations gives a relation between the cross sections which can be written as an expression for the $(\Omega^-, \overline{\Omega^-})$ production cross sections expressed in terms of the other three, namely

$$\sigma(\Omega^-\overline{\Omega^-}) = \sigma(N^{*-}\overline{N^{*-}}) + 3\{\sigma(\Xi^{*-}\overline{\Xi^{*-}}) - \sigma(Y^{*-}\overline{Y^{*-}})\}\,. \tag{C.17}$$

This is the only relation between these cross sections which is expressible as an equality. However, it is also possible to obtain relations expressible as inequalities. Such inequalities were of particular interest for Ω^- production processes since they can be expressed as a lower limit on the production cross section for reactions which have not yet been observed.

The amplitude for Ω^- production can be expressed in terms of the amplitude for any two of the other three processes (1), by eliminating a_0 and a_1.

$$
\begin{aligned}
(\bar{p}p|\Omega^-\overline{\Omega^-}) &= 2(\bar{p}p|N^{*-}\overline{N^{*-}}) + 3(\bar{p}p|Y^{*-}\overline{Y^{*-}})\,, && \text{(C.18a)}\\
&= \tfrac{1}{2}(\bar{p}p|N^{*-}\overline{N^{*-}}) - \tfrac{3}{2}(\bar{p}p|\Xi^{*-}\overline{\Xi^{*-}})\,, && \text{(C.18b)}\\
&= -(\bar{p}p|Y^{*-}\overline{Y^{*-}}) - 2(\bar{p}p|\Xi^{*-}\overline{\Xi^{*-}})\,. && \text{(C.18c)}
\end{aligned}
$$

Using the triangular inequality we obtain lower limits for the magnitude of the Ω^- production matrix element.

$$
\begin{aligned}
|(\bar{p}p|\Omega^-\overline{\Omega^-})| &\geqslant \big|2|(\bar{p}p|N^{*-}\overline{N^{*-}})| - 3|(\bar{p}p|Y^{*-}\overline{Y^{*-}})|\big| && \text{(C.19a)}\\
&\geqslant \big|\tfrac{1}{2}|(\bar{p}p|N^{*-}\overline{N^{*-}})| - \tfrac{3}{2}|(\bar{p}p|\Xi^{*-}\overline{\Xi^{*-}})|\big| && \text{(C.19b)}\\
&\geqslant \big|\,|(\bar{p}p|Y^{*-}\overline{Y^{*-}})| - 2|(\bar{p}p|\Xi^{*-}\overline{\Xi^{*-}})|\big|. && \text{(C.19c)}
\end{aligned}
$$

3. *Selection rules.* We consider first the decay of a vector meson which is a (0,0) singlet into two pseudoscalar mesons in (1,1) octets. We have already seen that parity conservation requires that the two

pseudoscalar mesons produced be in an odd-parity state and therefore in an SU_3 multiplet which is antisymmetric under permutation of the two octets. Since the (0,0) singlet is symmetric, the decay of a singlet vector meson into two pseudoscalar mesons is forbidden by SU_3. The same selection rule is obtainable using U-spin and isospin. A (0,0) singlet vector meson is a pure singlet in both isospin and U-spin having $T=U=0$. However, an odd-parity state of two pions must be antisymmetric in isospin and have $T=1$. The K^0 and $\overline{K}^0$ both belong to the same U-spin triplet and the generalized Pauli principle for U-spin requires that an odd-parity state be antisymmetric in U-spin; i.e. have $U=1$. The 2π and 2K decay modes are thus forbidden by isospin and U-spin conservation respectively. A simple generalization of this selection rule states that any boson of odd parity which is either in a (0,0) singlet or the 27-dimensional (2,2) multiplet cannot decay into two pseudoscalar mesons in (1,1) octets. Similarly, a boson of even parity which is in a (3,0) or (0,3) decuplet cannot decay into two (1,1) pseudoscalar mesons. Two pseudoscalar mesons in a symmetric (0,0) or (2,2) state must have even parity while if they are in the antisymmetric (3,0) or (0,3) state they must have odd parity.

C. The symmetry-breaking interactions

Unitary symmetry is broken by at least the following two interactions (i) the electromagnetic interaction and (ii) the part of the strong interaction which produces the mass splittings within the SU_3 multiplets. Experimental predictions for processes involving the symmetry-breaking interactions can be made using the transformation properties of these interactions under SU_3. These are conveniently expressed in terms of isospin and U-spin. Since isospin is known experimentally to be conserved in strong interactions, the part of the strong interaction which breaks SU_3 symmetry must still conserve isospin. The electromagnetic interaction, on the other hand, conserves U-spin. This can be seen as follows: The U-spin operators U_+ and U_- change the strangeness of a particle but not its electric charge. U-spin multiplets thus consist of particles all having the same coupling to the electromagnetic field. The electromagnetic interaction therefore conserves U-spin rigorously to all orders and the photon can be considered to be a particle of zero U-spin. The addition of strong interactions produces anomalous magnetic moments and affects the coupling of particles to the electromagnetic field. However, if strong interactions are invariant under SU_3, they also conserve U-spin. Thus U-spin is conserved in any combination of electromagnetic interactions and strong interactions invariant under SU_3.

The electromagnetic interaction which conserves U-spin does not conserve isospin and the mass-splitting interaction which conserves isospin does not conserve U-spin. If either interaction conserved both isospin and U-spin it would be invariant under SU_3 and would not break the symmetry. Let us now specify more precisely the isospin transformation properties of the electromagnetic interaction and the U-spin transformation properties of the mass-splitting interaction. The electromagnetic interaction is a linear combination of an isoscalar and an isovector. The mass-splitting interaction is assumed to be a linear combination of a U-spin scalar and a U-spin vector. These transformation properties are summarized in Table C.1.

TABLE C.1

Transformation properties of several interactions
S = scalar; V = vector

Interaction	*Isospin*	*U-spin*
Invariant under SU_3	S (conserved)	S (conserved)
Electromagnetic	S + V	S (conserved)
M	S (conserved)	S + V

The linear combinations of scalar and vector can be specified more precisely. The electromagnetic interaction transforms like the electric charge operator Q which is a member of a (1, 1) octet. The mass-splitting interaction transforms like the hypercharge operator Y which is also a member of a (1, 1) octet. That these interactions transform like members of an octet is also expressible in terms of isospin and U-spin. Using the octet transformation coefficients

(B.11) and (B.12) the electromagnetic interaction is a U-spin scalar which is a linear combination of an isoscalar and an isovector. The isoscalar and isovector parts are each linear combinations of a U-spin scalar and a U-spin vector, and the isospin and U-spin scalars and vectors are related by the octet transformation coefficients. Thus if we let U_s, U_v, I_s and I_v represent U-spin scalars and vectors we have

$$E \equiv U_s = \tfrac{1}{2}\sqrt{3}I_v \quad -\tfrac{1}{2}I_s \; , \qquad \text{(C.20a)}$$

$$U_v = \quad \tfrac{1}{2}I_v \; + \tfrac{1}{2}\sqrt{3}\, I_s \; , \qquad \text{(C.20b)}$$

$$M \equiv \; I'_s = \tfrac{1}{2}\sqrt{3}U'_v \quad -\tfrac{1}{2}\, U'_s, \qquad \text{(C.21a)}$$

$$I'_v = \quad \tfrac{1}{2}U'_v + \tfrac{1}{2}\sqrt{3}\; U'_s. \qquad \text{(C.21b)}$$

These relations are the same as the ones defining those particular linear combinations of the isoscalar Λ and the isovector Σ^0 which are a U-spin scalar and a U-spin vector.

There are two types of predictions of experimental properties which can be made for each symmetry-breaking interaction: (1) those following from the conserved spin, and (2) those following from the specific transformation property under the non-conserved spin. Results following from U-spin conservation in electromagnetic interactions are valid to all orders in the electromagnetic interaction and in the strong interactions invariant under SU_3. Results following from the specific transformation property of the electromagnetic interaction (C.20) are valid to all orders in strong interactions invariant under SU_3 but each order in the electromagnetic interaction must be considered separately. Results following from isospin conservation are good all to orders in the strong interactions including both those invariant under SU_3 and the mass-splitting interaction. Results following from the specific transformation property of the mass-splitting interaction (C.21) under U-spin are good to all orders in the interactions invariant under SU_3 but must be considered separately for each order in the mass-splitting interaction. Relations following from isospin conservation are valid independently of SU_3 and are well known. These will not be considered further. Let us now consider some examples of the other three types of experimental predictions.

1. *Relations following from U-spin conservation in electromagnetic interactions*

a) *Electromagnetic decays of the η and π^0.* Since photons have $U=0$, only a $U=0$ state can decay into two photons. The linear combination $|\frac{1}{2}\pi^0+\frac{1}{2}\sqrt{3}\eta\rangle$ is a pure U-spin triplet. Thus U-spin conservation requires the vanishing of the transition matrix element $\langle\frac{1}{2}\pi^0+\frac{1}{2}\sqrt{3}\eta|2\gamma\rangle=0$. This can be expressed as a relation between the transition matrix elements for the π^0- and η-decays

$$\langle\pi^0|2\gamma\rangle = -\sqrt{3}\langle\eta|2\gamma\rangle\,. \tag{C.22}$$

b) *Electromagnetic decays of the* N*, Y*, *and* Ξ*. The Y_1^{*-} and Ξ^{*-} belong to a $U=\frac{3}{2}$ quartet, while the Y_1^{*+} and N^{*+} belong to a $U=\frac{1}{2}$ doublet. The Σ^-, Ξ^-, Σ^+ and p all have $U=\frac{1}{2}$. Thus we find that the electromagnetic decays of the Y^{*-} and Ξ^{*-} are forbidden but the decays of the Y^{*+} and N^{*+} are allowed and have equal amplitudes, i.e.,

Forbidden: $\langle Y_1^{*-}|\Sigma^-\gamma\rangle$ and $\langle\Xi^{*-}|\Xi^-\gamma\rangle$

Allowed: $\langle Y_1^{*+}|\Sigma^+\gamma\rangle = \langle N^{*+}|p\gamma\rangle\,.$

The N^{*0}, Ξ^{*0} and Y^{*0} belong to the same $U=1$ triplet and can decay electromagnetically to the corresponding members of the $U=1$ baryon triplet consisting of the n, the Ξ^0, and the linear combination $\frac{1}{2}\Sigma^0+\frac{1}{2}\sqrt{3}\Lambda$. Thus

$$\langle N^{*0}|n\gamma\rangle = \langle\Xi^{*0}|\Xi^0\gamma\rangle = 2\langle Y^{*0}|\Sigma^0\gamma\rangle = \tfrac{2}{3}\sqrt{3}\langle Y^{*0}|\Lambda\gamma\rangle\,. \tag{C.23}$$

c) *Photoproduction of* N* *and* Y*. Consider the two reactions:

$$\begin{array}{lcccc} & \gamma+p & \to & N^{*0}+\pi^+\,, & \\ & \gamma+p & \to & Y^{*0}+K^+\,, & \\ U: & 0\ \ \tfrac{1}{2} & & 1\ \ \ \ \tfrac{1}{2}\,, & \\ U_{\text{total}}: & \tfrac{1}{2} & & \tfrac{1}{2}\text{ or }\tfrac{3}{2}\,. & \end{array}$$

The (γ, p) system is a U-spin eigenstate with $U=\frac{1}{2}$ since the photon has $U=0$ and the proton has $U=\frac{1}{2}$. The N^{*0} and the Y^{*0} are members of a $U=1$ triplet. The π^+ and K^+ are members of a $U=\frac{1}{2}$ doublet. If U-spin is conserved in these reactions, there is only one possible U-spin channel, namely $U=\frac{1}{2}$. The branching ratio for the two reactions thus depends only on the Clebsch–Gordan coefficients describing the coupling of a spin 1 and a spin $\frac{1}{2}$ to a total spin $\frac{1}{2}$. It is

$$\frac{\langle \gamma p | N^{*0}\pi^{+}\rangle}{\langle \gamma p | Y^{*0}K^{+}\rangle} = \frac{(1\tfrac{1}{2}1 -\tfrac{1}{2}|\tfrac{1}{2}\tfrac{1}{2})}{(1\tfrac{1}{2}0\ \ \tfrac{1}{2}|\tfrac{1}{2}\tfrac{1}{2})} = -\sqrt{3}\,. \qquad \text{(C.24)}$$

d) *Meson photoproduction.* Consider the reactions

$$\begin{array}{lllll} & \gamma + p \rightarrow & n & + \pi^{+} & , \\ & \gamma + p \rightarrow & \Lambda & + K^{+} & , \\ & \gamma + p \rightarrow & \Sigma^{0} & + K^{+} & , \\ U: & 0\ \ \tfrac{1}{2} & 1 \text{ or } 0 & \tfrac{1}{2} & , \\ U_{\text{total}}: & \tfrac{1}{2} & \tfrac{1}{2} \text{ or } \tfrac{3}{2}. & & \end{array}$$

Here again, the left-hand side has $U = \frac{1}{2}$ and is a pure U-spin eigenstate. The situation on the right-hand side of these reactions is more complicated because the Λ and Σ^0 are not U-spin eigenstates but are mixtures of $U=0$ and $U=1$. These particles can combine in two ways to make a $U=\frac{1}{2}$ state. Either the $U=0$ or the $U=1$ component of the Λ and Σ^0 can be coupled to the $U=\frac{1}{2}$ meson to obtain a total U-spin of $\frac{1}{2}$. There are therefore two independent complex amplitudes describing these three reactions. The existence of two complex amplitudes implies three real parameters: two magnitudes and one relative phase. Since there are only three cross sections, one cannot relate these *cross sections* by an equality. However, the relation between the *amplitudes* for the three reactions can be obtained and leads to inequalities relating the cross sections. These relations are obtained most easily by noting that the linear combination $\frac{1}{2}\Sigma^0 + \frac{1}{2}\sqrt{3}\Lambda$ is a U-spin eigenstate and belongs to the same $U=1$ triplet as the neutron. The amplitudes for the photoproduction of this particular linear combination and for neutron production thus are related by the Clebsch–Gordan coefficient describing the coupling of a spin 1 with a spin $\frac{1}{2}$ to a total spin $\frac{1}{2}$. The ratio of amplitudes therefore is

$$\frac{\langle \gamma p | n\pi^{+}\rangle}{\langle \gamma p | \{\tfrac{1}{2}\Sigma^{0} + \tfrac{1}{2}\sqrt{3}\Lambda\}K^{+}\rangle} = -\sqrt{2} \qquad \text{(C.25)}$$

so that

$$\langle \gamma p | n\pi^{+}\rangle = -\tfrac{1}{2}\sqrt{2}\langle \gamma p | \Sigma^{0}K^{+}\rangle - \tfrac{1}{2}\sqrt{6}\langle \gamma p | \Lambda K^{+}\rangle\,. \qquad \text{(C.26)}$$

This relation between *amplitudes* is not a relation between *cross sections* because the relative phase of the Σ^0 and Λ production amplitudes is unknown. However, it follows that the absolute values of the transition matrix elements are related by the inequalities

$$|\langle\gamma p|n\pi^+\rangle| \leqslant \tfrac{1}{2}\sqrt{2}|\langle\gamma p|\Sigma^0 K^+\rangle| + \tfrac{1}{2}\sqrt{6}|\langle\gamma p|\Lambda K^+\rangle|, \quad \text{(C.27a)}$$

$$|\langle\gamma p|n\pi^+\rangle| \geqslant \left|\tfrac{1}{2}\sqrt{2}|\langle\gamma p|\Sigma^0 K^+\rangle| - \tfrac{1}{2}\sqrt{6}|\langle\gamma p|\Lambda K^+\rangle|\right|. \quad \text{(C.27b)}$$

e) *Two-meson photoproduction.* U-spin conservation is particularly useful in considering reactions in which several particles are produced in the final state. If SU_3 algebra is used and all possible final states are considered, there are many possible couplings and many channels. Most relations obtained in this way are complicated. U-spin provides a method of choosing the particular reactions having simple properties; i.e. those reactions for which only one or two couplings are allowed by U-spin conservation. Consider for example the reactions

$$\begin{array}{lccccccccc} & \gamma & + & p & \rightarrow & N^{*-} & + & \pi^+ & + & \pi^+ , \\ & \gamma & + & p & \rightarrow & Y^{*-} & + & K^+ & + & \pi^+ , \\ & \gamma & + & p & \rightarrow & \Xi^{*-} & + & K^+ & + & K^+ . \\ U: & 0 & & \tfrac{1}{2} & & \tfrac{3}{2} & & \tfrac{1}{2} & & \tfrac{1}{2} \end{array}$$

U-spin conservation requires that the two mesons should couple to $U=1$ in order that the combination may couple with the $U=\frac{3}{2}$ baryon resonances to a total $U=\frac{1}{2}$. Thus

$$\begin{aligned} &\langle\gamma p|N^{*-}\pi^+\pi^+\rangle/\langle\gamma p|Y^{*-}\tfrac{1}{2}\sqrt{2}\{K^+\pi^+ + \pi^+K^+\}\rangle/\langle\gamma p|\Xi^{*-}K^+K^+\rangle \\ &\quad = (\tfrac{3}{2}\,1\,\tfrac{3}{2}\,-1|\tfrac{1}{2}\,\tfrac{1}{2})/(\tfrac{3}{2}\,1\,\tfrac{1}{2}\,0|\tfrac{1}{2}\,\tfrac{1}{2})/(\tfrac{3}{2}\,1\,-\tfrac{1}{2}\,1|\tfrac{1}{2}\,\tfrac{1}{2}) \\ &\quad = \quad 1/\sqrt{2} \quad / \quad -1/\sqrt{3} \quad / \quad 1/\sqrt{6}\,. \end{aligned} \quad \text{(C.28)}$$

The K^+ and π^+ are spinless bosons belonging to the same U-spin doublet. The wave function for a two-particle $K^+\pi^+$ system must be totally symmetric in space and U-spin. Since the $U=1$ state, $\frac{1}{2}\sqrt{2}\{K^+\pi^+ + \pi^+K^+\}$ is symmetric in U-spin, it must also be space-symmetric. The angular distribution for this reaction is therefore symmetric under interchange of the two mesons. They must be in a state of even orbital angular momentum and even parity with respect to their center of mass.

Similar relations follow for the production of the corresponding vector mesons.

f) *Static electromagnetic properties.* Any electromagnetic operator E is a U-spin scalar. Expectation values of such an operator in states of the same U-spin multiplet are therefore all equal.

Consider for example the baryon octet. The proton and the Σ^+ are members of the same U-spin doublet, and similarly for the Σ^- and Ξ^-. The neutron and Ξ^0 are members of the same U-spin triplet. Therefore for any electromagnetic operator E we have

$$\langle \mathrm{p}|E|\mathrm{p}\rangle = \langle \Sigma^+|E|\Sigma^+\rangle \,, \tag{C.29a}$$

$$\langle \Sigma^-|E|\Sigma^-\rangle = \langle \Xi^-|E|\Xi^-\rangle \,, \tag{C.29b}$$

$$\langle \mathrm{n}|E|\mathrm{n}\rangle = \langle \Xi^0|E|\Xi^0\rangle \,. \tag{C.29c}$$

If E represents the magnetic-moment operator, then eqs. (C.29) predict that the magnetic moments of the proton and Σ^+ are equal, and similarly for the Σ^- and Ξ^- and for the neutron and the Ξ^0. On the other hand if E is interpreted to be the electromagnetic mass-splitting operator, eqs. (C.29) can be combined to obtain a relation between the mass splittings in the three isotopic multiplets, namely

$$m_{\mathrm{p}} - m_{\mathrm{n}} = m_{\Sigma^+} - m_{\Sigma^-} + m_{\Xi^-} - m_{\Xi^0} \,. \tag{C.30}$$

2. *Relations depending upon the specific transformation properties of the electromagnetic interaction*

a) *Electromagnetic properties of the Λ and Σ^0.* U-spin conservation alone is not sufficient to obtain relations between the electromagnetic properties of the Λ and Σ^0 and the other members of the baryon octet, since these neutral baryons in the center of the octet are not U-spin eigenstates but are mixtures of $U=0$ and $U=1$. Further relations involving the electromagnetic properties of these particles can be obtained by using the specific transformation properties (C.20) of any electromagnetic operator under isospin and U-spin.

We first note that the expectation value of any isovector vanishes for both the Λ and the Σ^0; i.e. in both $|T=0\rangle$ and in $|T=1, T_0=0\rangle$ states:

$$\langle \Lambda|I_{\mathrm{v}}|\Lambda\rangle = \langle \Sigma^0|I_{\mathrm{v}}|\Sigma^0\rangle = 0 \,. \tag{C.31}$$

Then, from (C.20),

$$\langle \Lambda|E|\Lambda\rangle = -\tfrac{1}{2}\langle \Lambda|I_s|\Lambda\rangle\,, \tag{C.32a}$$

$$\langle \Sigma^0|E|\Sigma^0\rangle = -\tfrac{1}{2}\langle \Sigma^0|I_s|\Sigma^0\rangle\,. \tag{C.32b}$$

Substituting for I_s from (C.20) into (C.32) we obtain

$$\langle \Lambda|E|\Lambda\rangle = -\tfrac{1}{4}\sqrt{3}\langle \Lambda|U_v|\Lambda\rangle + \tfrac{1}{4}\langle \Lambda|U_s|\Lambda\rangle\,, \tag{C.33a}$$

$$\langle \Sigma^0|E|\Sigma^0\rangle = -\tfrac{1}{4}\sqrt{3}\langle \Sigma^0|U_v|\Sigma^0\rangle + \tfrac{1}{4}\langle \Lambda|U_s|\Lambda\rangle\,. \tag{C.33b}$$

But U_s appearing on the right-hand side of eqs. (C.33) is just equal to E by eq. (C.20a). Thus

$$\langle \Lambda|E|\Lambda\rangle = -\tfrac{1}{3}\sqrt{3}\langle \Lambda|U_v|\Lambda\rangle\,, \tag{C.34a}$$

$$\langle \Sigma^0|E|\Sigma^0\rangle = -\tfrac{1}{3}\sqrt{3}\langle \Sigma^0|U_v|\Sigma^0\rangle\,. \tag{C.34b}$$

The Λ and Σ^0 are both linear combinations of a state $|U=0\rangle$ and a state $|U=1\rangle$

$$|\Lambda\rangle = \tfrac{1}{2}\sqrt{3}|U=1\rangle - \tfrac{1}{2}|U=0\rangle\,. \tag{C.35a}$$

$$|\Sigma^0\rangle = \tfrac{1}{2}|U=1\rangle + \tfrac{1}{2}\sqrt{3}|U=0\rangle\,. \tag{C.35b}$$

Eq. (C.34) simplifies when the Λ and Σ^0 are expressed in terms of the U-spin eigenstates, eq. (C.35), because any U-spin vector has a vanishing diagonal matrix element in both the $|U=0\rangle$ and the $|U=1, U_0=0\rangle$ states. The only non-vanishing matrix element of U_v is the off-diagonal element. Thus

$$\langle \Lambda|U_v|\Lambda\rangle = -\tfrac{1}{2}\sqrt{3}\,\mathrm{Re}\,\langle U=1|U_v|U=0\rangle\,, \tag{C.36a}$$

$$\langle \Sigma^0|U_v|\Sigma^0\rangle = +\tfrac{1}{2}\sqrt{3}\,\mathrm{Re}\,\langle U=1|U_v|U=0\rangle\,. \tag{C.36b}$$

Substituting eq. (C.36) into eq. (C.34) we obtain

$$\langle \Lambda|E|\Lambda\rangle = -\langle \Sigma^0|E|\Sigma^0\rangle\,. \tag{C.37}$$

If E is the magnetic moment operator we find that the magnetic moments of the Λ and Σ^0 are equal and opposite. On the other hand, we can also calculate the expectation value of E in the states (C.35) using the fact that E is a U-spin scalar

$$\langle \Lambda|E|\Lambda\rangle = \tfrac{3}{4}\langle U=1|U_s|U=1\rangle + \tfrac{1}{4}\langle U=0|U_s|U=0\rangle\,, \tag{C.38a}$$

$$\langle \Sigma^0|E|\Sigma^0\rangle = \tfrac{1}{4}\langle U=1|U_s|U=1\rangle + \tfrac{3}{4}\langle U=0|U_s|U=0\rangle\,. \tag{C.38b}$$

The sum of the two equations (C.38) must vanish by eq. (C.37). Thus

$$\langle U=1|U_s|U=1\rangle = -\langle U=0|U_s|U=0\rangle = 2\langle \Lambda|E|\Lambda\rangle\,. \tag{C.39}$$

Since the neutron is in the same U-spin triplet as the $U=1$ linear combination of the Λ and Σ^0,

$$\langle \mathrm{n}|E|\mathrm{n}\rangle = \langle U=1|U_s|U=1\rangle = 2\langle\Lambda|E|\Lambda\rangle = -2\langle\Sigma^0|E|\Sigma^0\rangle\,. \quad \text{(C.40)}$$

Thus if E is the magnetic moment operator, the magnetic moment of the neutron is equal to twice the magnetic moment of the Λ. The $\langle\Lambda|E|\Sigma\rangle$ transition matrix element can also be obtained

$$\begin{aligned}\langle\Lambda|E|\Sigma\rangle &= \tfrac{1}{4}\sqrt{3}\langle U=1|U_s|U=1\rangle - \tfrac{1}{4}\sqrt{3}\langle U=0|U_s|U=0\rangle \\ &= \sqrt{3}\langle\Lambda|E|\Lambda\rangle\,.\end{aligned} \quad \text{(C.41)}$$

b) *Electromagnetic decays of vector mesons* (An interesting exercise in symmetries). Okubo has given the following relations between the decays of a vector meson into a pseudoscalar meson and a photon, assuming invariance of strong interactions under the transformations of SU_3 and under charge conjugation

$$\begin{aligned}\langle\varrho^+|\pi^+\gamma\rangle &= \langle\varrho^0|\pi^0\gamma\rangle = \langle K^{*+}|K^+\gamma\rangle = -\langle\omega_8|\eta\gamma\rangle \\ &= \tfrac{1}{3}\sqrt{3}\langle\omega_8|\pi^0\gamma\rangle = \tfrac{1}{3}\sqrt{3}\langle\varrho_0|\eta\gamma\rangle = -\tfrac{1}{2}\langle K^{0*}|K^0\gamma\rangle\,. \end{aligned} \quad \text{(C.42)}$$

These relations are valid to first order in the electromagnetic interaction, to all orders in strong interactions invariant under SU_3, and do not take into account the part of the strong interaction which breaks unitary symmetry.

A detailed derivation of Okubo's relation using isospin and U-spin as well as charge conjugation is instructive because it illustrates the interplay of the three symmetries. One finds that Okubo's relation is actually a combination of several groups of relations, each having a wider range of validity than the combined relation (C.42).

The following properties of the symmetry-breaking interactions (Table C.1) are relevant here:

(1) The symmetry-breaking part of the strong interactions conserves isospin; it does not conserve U-spin.

(2) The electromagnetic interaction conserves U-spin; it does not conserve isospin.

(3) The electromagnetic interaction consists of an isoscalar and an isovector part. Selection rules imposed by U-spin conservation are good to all orders in the electromagnetic interaction. Selection rules following from isospin transformation properties must be

defined specifically for each order in the electromagnetic interaction.
(4) The electromagnetic interaction transforms under SU_3 like the member of the octet representation, which is a U-spin scalar. This is expressed in terms of isospin and U-spin by the relations (C.28). Again, relations requiring only that the electromagnetic interaction be a U-spin scalar are valid to all orders in the electromagnetic interaction while those using the explicit transformation properties (C.20) must be considered separately for each order.

Using these properties, three sets of relations can be obtained which, when combined, lead to Okubo's result (C.42).
(1) Relations following from isospin and charge conjugation. Because of the mixed isospin of the photon, simple relations do not follow easily from isospin conservation alone. The combination of isospin and charge conjugation is conveniently expressed by defining a G-parity for the photon. Since it is odd under charge conjugation, it follows that the isoscalar part has negative G-parity and the isovector part has positive G-parity. Thus, if all the strongly interacting particles participating in a given electromagnetic reaction have a definite G-parity, conservation of G-parity eliminates either the isoscalar or the isovector part of the photon and isospin conservation can then be used in the conventional manner.

The ϱ-meson has positive G-parity; the π has negative G-parity. G-conservation then requires negative G-parity for the photon emitted in the decay of a ϱ-meson into a π and a photon; i.e. only the isoscalar part contributes. The photon in these decays can therefore be considered as a scalar under isospin. Thus

$$\langle \varrho^+ | \pi^+ \gamma \rangle = \langle \varrho^0 | \pi^0 \gamma \rangle = \langle \varrho^- | \pi^- \gamma \rangle \,. \qquad \text{(C.43)}$$

This result is valid to all orders in all strong interactions conserving isospin (including the part of the strong interaction which breaks unitary symmetry) but is valid only to first order in the electromagnetic interaction.

One second-order electromagnetic process would be the emission of two photons. Here again a negative G-parity is required for the two-photon system and can be obtained only by taking the isoscalar part of one photon and the isovector part of the other.

This leads to a unique isospin transformation property for the two-photon system, namely pure isovector. Isospin conservation then leads to the relations

$$\langle \varrho^+ | \pi^+ 2\gamma \rangle = -\langle \varrho^- | \pi^- 2\gamma \rangle \,, \quad \text{(C.44a)}$$

$$\langle \varrho^0 | \pi^0 2\gamma \rangle = 0 \,. \quad \text{(C.44b)}$$

Inspection of the relations (C.44) shows that these follow from charge conjugation alone and are independent of isospin conservation. Only the first-order relation (C.43) gives an additional restriction imposed by isospin conservation over that obtained from charge conjugation alone, namely the relation between the neutral decay to the charged decay. Isospin conservation becomes completely useless in higher-order processes since G-parity always allows several values for the isospin of the photon system and these constitute independent isospin channels.

(2) Relations following from U-spin conservation.
Since the photon is a scalar under U-spin transformations, it follows immediately from U-spin conservation that

$$\langle \varrho^+ | \pi^+ n\gamma \rangle = \langle K^{*+} | K^+ n\gamma \rangle \,, \quad \text{(C.45a)}$$

$$\langle \varrho^- | \pi^- n\gamma \rangle = \langle K^{*-} | K^- n\gamma \rangle \,, \quad \text{(C.45b)}$$

$$\langle K^{*0} | K^0 n\gamma \rangle = \langle \overline{K^{*0}} | \overline{K^0} n\gamma \rangle \,. \quad \text{(C.45c)}$$

These relations are valid for the emission of any number n of photons to all orders in the electromagnetic interaction and in strong interactions invariant under SU_3. They do not take into account the symmetry-breaking part of the strong interaction, which is not invariant under SU_3. Further relations of this type are also obtainable for the neutral members of the meson octets. These are obtained simply by expressing these neutral mesons in terms of U-spin eigenstates. Let V_1, V_0, P_1 and P_0 represent vector and pseudoscalar mesons with U-spin eigenvalues of 1 and 0, respectively. Then

$$\varrho^0 = \tfrac{1}{2}V_1 + \tfrac{1}{2}\sqrt{3}\,V_0 \,, \quad \text{(C.46a)}$$

$$\pi^0 = \tfrac{1}{2}P_1 + \tfrac{1}{2}\sqrt{3}\,P_0 \,, \quad \text{(C.46b)}$$

$$\omega_8 = \tfrac{1}{2}\sqrt{3}\,V_1 - \tfrac{1}{2}V_0 \,, \quad \text{(C.46c)}$$

$$\eta = \tfrac{1}{2}\sqrt{3}\,P_1 - \tfrac{1}{2}P_0 \,. \quad \text{(C.46d)}$$

Since U-spin conservation forbids transitions between states of U-spin 1 and states of U-spin 0,

$$\langle K^{*0}|K^0\,n\gamma\rangle = \langle V_1|P_1\,n\gamma\rangle = \langle \overline{K^{*0}}\,|\,\overline{K^0}\,n\gamma\rangle\,, \tag{C.47a}$$

$$\langle \varrho^0|\eta\,n\gamma\rangle \quad = \tfrac{1}{4}\sqrt{3}\{\langle V_1|P_1 n\gamma\rangle - \langle V_0|P_0 n\gamma\rangle\} = \langle \omega_8|\pi^0 n\gamma\rangle, \tag{C.47b}$$

$$\langle \varrho^0|\pi^0\,n\gamma\rangle \quad = \tfrac{1}{4}\langle V_1|P_1\,n\gamma\rangle + \tfrac{3}{4}\langle V_0|P_0\,n\gamma\rangle\,, \tag{C.47c}$$

$$\langle \omega_8|\eta\,n\gamma\rangle \quad = \tfrac{3}{4}\langle V_1|P_1\,n\gamma\rangle + \tfrac{1}{4}\langle V_0|P_0\,n\gamma\rangle\,. \tag{C.47d}$$

Combining eqs. (C.47) gives

$$\langle \varrho^0|\eta\,n\gamma\rangle = \langle \omega_8|\pi^0\,n\gamma\rangle = -\tfrac{1}{2}\sqrt{3}\{\langle \varrho^0|\pi^0\,n\gamma\rangle - \langle \omega_8|\eta\,n\gamma\rangle\}\,, \tag{C.48a}$$

$$\langle \varrho^0|\pi^0\,n\gamma\rangle + \langle \omega_8|\eta\,n\gamma\rangle + \tfrac{4}{3}\sqrt{3}\langle \varrho^0|\eta\,n\gamma\rangle = 2\langle K^{*0}|K^0\,n\gamma\rangle\,, \tag{C.48b}$$

where again the relations hold for any arbitrary number n of photons since photons are U-spin scalars and can carry no U-spin *The result (C.48) is also good to all orders in the electromagnetic interaction and in strong interactions invariant under SU_3.*

(3) Relations requiring the specific transformation property of the photon under SU_3.

The relations (C.43), (C.45) and (C.48) are still insufficient to give Okubo's relation (C.42). One additional relation is required which is based upon both isospin and U-spin transformation properties for the photon. Using eqs. (C.20) and isospin conservation we have for a first-order electromagnetic process

$$\langle \varrho^0|E|\pi^0\rangle = -\tfrac{1}{2}\langle \varrho^0|I_s|\pi^0\rangle = -\tfrac{1}{4}\sqrt{3}\langle \varrho^0|U_v|\pi^0\rangle + \tfrac{1}{4}\langle \varrho^0|U_s|\pi^0\rangle\,, \tag{C.49a}$$

$$\langle \omega_8|E|\eta\rangle = -\tfrac{1}{2}\langle \omega_8|I_s|\eta\rangle = -\tfrac{1}{4}\sqrt{3}\langle \omega_8|U_v|\eta\rangle + \tfrac{1}{4}\langle \omega_8|U_s|\eta\rangle\,, \tag{C.49b}$$

since the isovector contribution to this matrix element vanishes. Since from eq. (C.20) $E = U_s$

$$\langle \varrho^0|E|\pi^0\rangle = -\tfrac{1}{3}\sqrt{3}\langle \varrho^0|U_v|\pi^0\rangle\,, \tag{C.50a}$$

$$\langle \omega_8|E|\eta\rangle = -\tfrac{1}{3}\sqrt{3}\langle \omega_8|U_v|\eta\rangle\,. \tag{C.50b}$$

Substituting the U-spin eigenstates (C.46)

$$\langle \varrho^0|E|\pi^0\rangle = -\tfrac{1}{4}\{\langle V_1|U_v|P_0\rangle + \langle P_1|U_v|P_0\rangle\} = -\langle \omega_8|E|\eta\rangle \tag{C.51}$$

or

$$\langle \varrho^0|\pi^0\,\gamma\rangle = -\langle \omega_8|\eta\gamma\rangle\,. \tag{C.52}$$

Eq. (C.52) is valid only to first order in the electromagnetic interac-

tion, to all orders in strong interactions invariant under SU_3, and neglects the symmetry-breaking part of the strong interactions.

Combining eqs. (C.43), (C.45), (C.48) and (C.52) leads to Okubo's relation (C.42). The validity of the relation (C.42) is limited by the conditions of validity of eq. (C.52) which are the most stringent.

3. *The mass-splitting interaction*

An outstanding success of the octet model of unitary symmetry has been the prediction of the mass splittings within SU_3 multiplets. It is assumed that these mass splittings are obtained by taking the expectation value of an operator having the transformation properties (C.21a). The proof of the pudding thus far seems to be in the eating; namely in the remarkable agreement between experiment and calculated predictions. The procedure is apparently one of using first-order degenerate perturbation theory to obtain a *large* effect with no indication that higher-order corrections should be small. There are presumably more profound reasons why such a procedure should work. However, these have not yet been established and are in any case beyond the scope of this treatment. We therefore consider only the problem of how the mass splittings can be calculated once this assumption is accepted. These are easily obtained using the isospin and U-spin transformation properties of the mass-splitting interaction (C.21). The results are assumed to give the splittings of the masses directly for fermions but the splitting in the values of the square of the mass for bosons because the mass enters linearly in fermion propagators and quadratically in boson propagators.

a) *The masses of the $\frac{3}{2}^+$ baryon resonances.* Consider the U-spin quartet of negatively charged baryons: Ω^-, Ξ^{*-}, Y_1^{*-} and N^{*-}. These have $U=\frac{3}{2}$ and $U_0=-\frac{3}{2}, -\frac{1}{2}, +\frac{1}{2}$ and $+\frac{3}{2}$, respectively. Let us consider the expectation value of the mass-splitting operator in this U-spin multiplet. The U-spin scalar part of the operator has the same expectation value for all states in the same U-spin multiplet and therefore does not give any mass splitting. The U-spin vector part gives a splitting which is proportional to U_0 within the same U-spin multiplet. Thus the mass splitting of the four members of the quartet is proportional to U_0 and the mass spacings are all equal; i.e. there are four equally spaced energy levels.

b) *The mass splitting in the baryon octet.* Consider the neutral members of the U-spin triplet $[\Xi^0, \frac{1}{2}(\Sigma^0+\sqrt{3}\Lambda), \mathrm{n}]$, which are states with $U=1$ and $U_0=-1$, 0 and $+1$, respectively. Here again the expectation value of the U-spin scalar part of the mass-splitting operator is the same for all three states. Let us denote this by S. The expectation value of the U-spin vector part of the mass-splitting operator is proportional to U_0. If the proportionality factor is denoted by V, the expectation value is therefore $-V$, 0 and $+V$, respectively, for the three states. We can therefore write

$$\langle \Xi^0 | M | \Xi^0 \rangle = S - V\,, \tag{C.53a}$$

$$\langle -\tfrac{1}{2}\Sigma^0 + \tfrac{1}{2}\sqrt{3}\Lambda | M | -\tfrac{1}{2}\Sigma^0 + \tfrac{1}{2}\sqrt{3}\Lambda \rangle = S\,, \tag{C.53b}$$

$$\langle \mathrm{n} | M | \mathrm{n} \rangle = S + V\,, \tag{C.53c}$$

where M represents the mass-splitting operator. Since M conserves isospin, it has no off-diagonal elements between the Λ and the Σ^0. Eq. (C.53b) can therefore be rewritten

$$\tfrac{1}{4}\langle \Sigma^0 | M | \Sigma^0 \rangle + \tfrac{3}{4}\langle \Lambda | M | \Lambda \rangle = S\,. \tag{C.53d}$$

Combining these equations leads to the result

$$\tfrac{1}{2}\langle \mathrm{n} | M | \mathrm{n} \rangle + \tfrac{1}{2}\langle \Xi^0 | M | \Xi^0 \rangle = \tfrac{1}{4}\langle \Sigma^0 | M | \Sigma^0 \rangle + \tfrac{3}{4}\langle \Lambda | M | \Lambda \rangle\,. \tag{C.54}$$

c) *The mass splitting in the meson octet.* Since the SU_3 couplings are the same for any octet, the results for the pseudoscalar meson octet are obtainable directly from the results for the baryon octet (C.54). However, as mentioned above, the results for bosons refer to the square of the mass rather than the mass itself, and the masses of the K and $\overline{\mathrm{K}}$ must be equal by charge conjugation. We thus obtain the result

$$M_{\mathrm{K}}^2 = \tfrac{1}{4}(M_\pi^2 + 3M_\eta^2) \tag{C.55}$$

where M_{K}, M_π and M_η are the masses of these three particles.

APPENDIX D

PHASES, A PERENNIAL HEADACHE

If one wishes to give a precise definition for the states within a multiplet which are transformed into one another by the action of the operators of a Lie algebra, there is a certain arbitrariness in the choice of phases. However, once a convention is chosen confusion and errors are avoided by using the same convention throughout a particular calculation*. It would be desirable to choose a phase convention** in which extra phase factors do not appear in the basic relations. This, however, is impossible as can be seen from the following simple example. Consider three operators A, B and C satisfying the following commutation relation

$$[A, B] = C\,. \tag{D.1}$$

It is then evident that

$$[B, A] = -C\,. \tag{D.2}$$

A negative sign must appear in one of the two commutation relations and cannot be avoided by any redefinition of the phases.

A particular example relevant to angular momentum algebra is

$$[J_z, J_+] = J_+, \qquad [J_+, J_z] = -J_+\,. \tag{D.3}$$

In these relations the angular momentum operators J_+ and J_z can be interpreted in two ways. On the one hand they are the operators of the angular momentum Lie algebra. These acting on any element of an angular momentum multiplet give another element of

* Do not believe this sentence. There are always confusion and errors. You have to live with them.

** The typist originally typed 'phase confusion' instead of phase convention. This error is too good to lose.

the multiplet. On the other hand the three angular momentum operators can also be considered as the members of a vector or triplet multiplet. The first of the two commutation relations (D.3) above can be interpreted as the action of the diagonal operator J_z on J_+ which is the plus element of a vector multiplet. Since the operator J_z is taken to be diagonal in the usual representation of angular momentum operators the plus state is an eigenstate of J_z with eigenvalue $+1$ as is indicated by the equation. The second of the two equations (D.3) can be interpreted like the first but with the roles of the operators J_z and J_+ reversed. We now have the operator J_+ as a step operator raising the eigenvalue of J_z within a multiplet. The operator J_z is now considered to be the zero element of the vector multiplet. The result of the operation is the plus element of the multiplet, namely J_+. However, a minus sign must appear in this relation if the previous relation (D.3) has a plus sign.

One therefore has the following choice to make when defining phase conventions. One can say that the three angular momentum operators J_+, J_z, and J_- represent the components of a vector without any additional phase factors. In that case a phase factor must appear in relations like (D.3) expressing the operation of the step operators. On the other hand one can also define a phase convention in which all the step operator relations have a positive sign. This is the convention adopted by Condon and Shortley, and which is in general use. However, for this case we see that J_+ and J_z cannot be defined as the plus and zero elements of a vector triplet. An extra minus sign must be introduced either in the definition of the zero element or the definition of the plus element in order to absorb the minus sign in the commutation relation (D.2).

The same difficulty arises in considering the components of any vector, such as the vector $\boldsymbol{r}$. One is tempted to take the Cartesian coordinates x, y and z of the vector and form the natural linear combinations $x+iy$, $x-iy$ and z. However, these transform under rotations exactly like J_+, J_- and J_z. The same difficulty in phases arises and a minus sign must be introduced somewhere.

This difficulty translated into isospin language has been the source of untold confusion in field theory. It is customary to start with three real pion fields π_1, π_2, and π_3. The charged pions are then designated by the linear combination $\pi_1+i\pi_2$ and $\pi_1-i\pi_2$. If the Condon and Shortley phase convention is to be used for step operator relations, a minus sign again has to be inserted somewhere. The result of this confusion is that if one wishes to describe a state of a many-pion system (such as a two-pion state with the pions coupled to isospin $T=1$) one can find any sign one pleases by looking at the appropriate place in the literature.

A second kind of difficulty arises in the consideration of particles and antiparticles, charge conjugation, and creation and annihilation operators. This is seen in its simplest form by examining the following operators describing a system of neutrons and protons. The operator

$$B = a_p^\dagger a_p + a_n^\dagger a_n \tag{D.4}$$

is clearly the number operator giving the total number of nucleons and is a scalar under isospin transformations. On the other hand, the operator

$$2\tau_0 = a_p^\dagger a_p - a_n^\dagger a_n \tag{D.5}$$

is the zero component of an isovector. The coupling of two spins of one-half using the Condon-Shortley phase convention places the positive sign in the triplet and the negative sign in the singlet, in disagreement with the expressions (D.4) and (D.5). It then follows that we cannot consider $(a_p^\dagger, a_n^\dagger)$ and (a_n, a_p) to be two isospin doublets. An extra negative sign is needed somewhere if Condon–Shortley phases are to be used. The customary choice is to make minus a_n the member of the isospin doublet rather than a_n. Since operators which annihilate particles also create the corresponding antiparticles the extra minus sign is also inserted in discussing the antinucleon doublet. The same situation then obtains in the case of the K-meson isospin doublet with a negative sign inserted by convention in the $\overline{K}^0$.

These same phase difficulties carry over into the discussion of all Lie groups. The first difficulty applies directly to the set of generators or the group of operators of the Lie algebra. These operators themselves constitute a multiplet, a vector in the case of angular momentum, and an octet (1, 1) in the case of SU_3. However, to define a multiplet from these operators consistent with the Condon and Shortley phase convention it is always necessary to introduce minus signs in certain places.

In the case of SU_3 further phase difficulties are encountered when attempting to treat the three SU_2 subgroups on an equivalent basis. One finds that it is impossible to define phases for a multiplet containing the generators in such a way that all of the step operators for the three SU_2 subgroups have a positive sign. This can be seen very simply by examining the Sakata model. The three SU_2 subgroups then correspond to transformations in the (n, p) space, the (n, Λ) space and the (Λ, p) space. Let us now consider the sakaton and antisakaton triplets. We have seen above that isospin operations on nucleons and antinucleons can be made consistent with the Condon and Shortley phase convention by adding a minus sign in the neutron annihilation (antineutron creation) operator. By symmetry it is evident that this negative sign also fixes up U-spin which operates in the (n, Λ) space. However, it does not help V-spin, which operates in the (Λ, p) space and is independent of the neutron and the phase of neutron operators. A phase must be changed either in the proton or Λ-operators to make V-spin step operator relations consistent with the Condon and Shortley phase convention. This then messes up either isospin or U-spin. It is impossible to change phases in such a way that one and only one minus sign appears in each of the following three pairs, (p, n), (n, Λ) and (Λ, p).

In this book no attempt has been made to choose a universal consistent set of phase conventions. In each specific example phases have been chosen for maximum convenience in that particular example. For all relations involving angular momentum algebra the Condon–Shortley phase convention has been used. In the treatment of SU_3 using isospin and U-spin, Condon and Shortley phases have been used, implying that these phases could not simultaneously be used for V-spin. Since V-spin is not used directly in these cases, the awkward V-spin phases pose no difficulty. Positive phases have been used for all particles in the treatments of elementary particles, both for isospin and U-spin, for reasons of simplicity. This leads to no errors in the results for physically measurable quantities. If amplitudes or matrix elements (rather than their absolute magnitudes) are used in conjunction with other results obtained elsewhere with different phase conventions, inconsistencies and errors arise. Note, for example, that positive phases are used for all isospin and U-spin triplets, whereas some negative phases are frequently used as discussed above for pions. Positive phases have been used for both particles and antiparticles in disagremeent with the argument given above for the nucleon and K-doublets. With the positive phases used here, negative signs would appear in awkward places in relations involving charge conjugation. However, the operation of charge conjugation is never employed explicitly in these examples, and the necessity for introducing the negative signs never occurs. However, inconsistencies can occur if these results, expressed as amplitudes, are combined with results obtained by other means using the standard phase convention.

BIBLIOGRAPHY

The purpose of this book is to present a new simpler approach to the application of Lie groups in physical problems, in order to make these techniques available to a large body of physicists who find the standard treatments unintelligible. Specific references to original works have not been given. These would naturally use the conventional methods and terminology of group theory and would not be useful to the average reader. A reader who is interested in looking more deeply into a particular topic needs a guide to the standard literature on this topic, rather than an original reference to a specific point. Such a guide is given below, and includes both general references on group theory and specific references to topics listed according to the chapter in which they appear in this book. The list is not meant to be complete, nor to contain the most significant works. It should be adequate to guide the reader on his own search through the literature.

Since references are not given, no attempt has been made to give proper credit for any of the original developments presented as examples of the group theoretical method. The use of proper names has been avoided wherever possible without rendering the material difficult to understand. Since the topics of strange particles, unitary symmetry and the eightfold way have been treated without mentioning the name of Gell-Mann, it is hoped that others who have made significant contributions in this field will not feel insulted at having their names omitted as well. A word of apology is perhaps in order for those whose names have somehow managed to find their way into the book, such as Lie, Young, Wigner, Eckart, Pauli, Clebsch, Gordan, Lorentz, Minkowsky, Elliott, Casimir, Condon, Shortley, Sakata and Okubo. It was harder to keep these names out without confusing the reader. We hope they will not be offended.

GENERAL REFERENCES ON GROUP THEORY

E. P. Wigner, Group Theory and its Application to the Quantum Mechanics of Atomic Spectra (English translation; Academic Press, New York 1958).

M. Hamermesh, Group Theory and its Application to Physical Problems (Addison-Wesley, Reading, Mass. 1962).

G. Racah, Group Theory and Spectroscopy, Lecture Notes (Princeton 1951).

A. Messiah, Quantum Mechanics **2** (English translation; North-Holland Publ. Co., Amsterdam 1963) Appendices C and D.

A. de-Shalit and I. Talmi, The Nuclear Shell Model (Academic Press, New York 1963).

CHAPTER 1, ANGULAR MOMENTUM

A. E. Edmonds, Angular Momentum in Quantum Mechanics (Princeton University Press, Princeton 1957).

M. E. Rose, Elementary Theory of Angular Momentum (Wiley, New York 1957).

U. Fano and G. Racah, Irreducible Tensorial Sets (Academic Press, New York 1959).

See also general references Messiah and de-Shalit and Talmi.

CHAPTER 2, ISOSPIN

J. M. Blatt and V. F. Weisskopf, Theoretical Nuclear Physics (Wiley, New York 1952) pp. 153, 220, 255.

M. G. Mayer and J. H. D. Jensen, Elementary Theory of Nuclear Shell Structure (Wiley, New York 1955) p. 158.

J. D. Jackson, The Physics of Elementary Particles (Princeton University Press, Princeton 1958) p. 5.

See also general references Messiah and de-Shalit and Talmi.

CHAPTER 3, SU_3 AND ELEMENTARY PARTICLES

M. Gell-Mann and Y. Ne'eman, The Eightfold Way (W. A. Benjamin, New York 1964).

Y. Ne'eman, Strong Interaction Symmetry. *In*: J. G. Wilson and S. A. Wouthuysen, eds., Progress in Elementary Particle and Cosmic Ray Physics **8** (North-Holland Publ. Co., Amsterdam, in press).

CHAPTER 4, THE THREE-DIMENSIONAL HARMONIC OSCILLATOR

J. P. Elliott, The Nuclear Shell Model and its Relation with Other Models. *In*: F. Janouch, ed., Selected Topics in Nuclear Theory (International Atomic Energy Agency, Vienna 1963).

CHAPTER 5, ALGEBRAS OF OPERATORS WHICH CHANGE THE NUMBER OF PARTICLES

Much of the material presented in this chapter is original and cannot be found in the literature at all. Seniority and properties of symplectic groups are treated in the general references Racah and de-Shalit and Talmi. Pairing quasispins were first introduced in the following references:

A. K. Kerman, Ann. Phys. (New York) **12** (1961) 300, for nuclear structure.

P. W. Anderson, Phys. Rev. **112** (1958) 164, for superconductivity.

A general review of the application of pairing quasispin methods to nuclear shell model calculations is given by:

R. D. Lawson and M. H. Macfarlane, submitted to Nuclear Physics.

The material presented in §§ 5.2, 5.5, 5.6 and 5.7 is original, much of it previously unpublished. The treatment of seniority with neutrons and protons using the group Sp_4 has now been undertaken by several investigators. The first published reference is:

K. Helmers, Nucl. Phys. **23** (1961) 594.

Unfortunately, no comprehensive review articles exist on these quasispin methods, and the pedestrian reader may find these original papers somewhat condensed and difficult reading.

The particular example mentioned in § 5.6 is treated in detail in the following paper:

S. Goshen and H. J. Lipkin, Ann. Phys. (New York) **6** (1959) 301.

An amusing example of the use of quasispin techniques (which might have been included in the book but wasn't) is found in the following reference:

H. J. Lipkin, Nucl. Phys. **26** (1961) 147.

CHAPTER 6, PERMUTATIONS, BOOKKEEPING AND YOUNG DIAGRAMS

This material is presented in detail in the general references on group theory listed above.

CHAPTER 7, THE GROUPS SU_4, SU_6 AND SU_{12}, AN INTRODUCTION TO GROUPS OF HIGHER RANK

F. J. Dyson, Symmetry Groups in Nuclear and Particle Physics (W. A. Benjamin, New York, 1966).

High Energy Physics and Elementary Particles (International Center for Theoretical Physics, Trieste; International Atomic Energy Agency, Vienna, 1965).

SUBJECT INDEX

A CATALOG OF SELECTED

DOVER BOOKS

IN SCIENCE AND MATHEMATICS

Astronomy

BURNHAM'S CELESTIAL HANDBOOK, Robert Burnham, Jr. Thorough guide to the stars beyond our solar system. Exhaustive treatment. Alphabetical by constellation: Andromeda to Cetus in Vol. 1; Chamaeleon to Orion in Vol. 2; and Pavo to Vulpecula in Vol. 3. Hundreds of illustrations. Index in Vol. 3. 2,000pp. 6⅛ x 9¼.
Vol. I: 0-486-23567-X
Vol. II: 0-486-23568-8
Vol. III: 0-486-23673-0

EXPLORING THE MOON THROUGH BINOCULARS AND SMALL TELESCOPES, Ernest H. Cherrington, Jr. Informative, profusely illustrated guide to locating and identifying craters, rills, seas, mountains, other lunar features. Newly revised and updated with special section of new photos. Over 100 photos and diagrams. 240pp. 8¼ x 11. 0-486-24491-1

THE EXTRATERRESTRIAL LIFE DEBATE, 1750–1900, Michael J. Crowe. First detailed, scholarly study in English of the many ideas that developed from 1750 to 1900 regarding the existence of intelligent extraterrestrial life. Examines ideas of Kant, Herschel, Voltaire, Percival Lowell, many other scientists and thinkers. 16 illustrations. 704pp. 5⅜ x 8½. 0-486-40675-X

THEORIES OF THE WORLD FROM ANTIQUITY TO THE COPERNICAN REVOLUTION, Michael J. Crowe. Newly revised edition of an accessible, enlightening book recreates the change from an earth-centered to a sun-centered conception of the solar system. 242pp. 5⅜ x 8½. 0-486-41444-2

A HISTORY OF ASTRONOMY, A. Pannekoek. Well-balanced, carefully reasoned study covers such topics as Ptolemaic theory, work of Copernicus, Kepler, Newton, Eddington's work on stars, much more. Illustrated. References. 521pp. 5⅜ x 8½.
0-486-65994-1

A COMPLETE MANUAL OF AMATEUR ASTRONOMY: TOOLS AND TECHNIQUES FOR ASTRONOMICAL OBSERVATIONS, P. Clay Sherrod with Thomas L. Koed. Concise, highly readable book discusses: selecting, setting up and maintaining a telescope; amateur studies of the sun; lunar topography and occultations; observations of Mars, Jupiter, Saturn, the minor planets and the stars; an introduction to photoelectric photometry; more. 1981 ed. 124 figures. 25 halftones. 37 tables. 335pp. 6½ x 9¼. 0-486-40675-X

AMATEUR ASTRONOMER'S HANDBOOK, J. B. Sidgwick. Timeless, comprehensive coverage of telescopes, mirrors, lenses, mountings, telescope drives, micrometers, spectroscopes, more. 189 illustrations. 576pp. 5⅝ x 8¼. (Available in U.S. only.)
0-486-24034-7

STARS AND RELATIVITY, Ya. B. Zel'dovich and I. D. Novikov. Vol. 1 of *Relativistic Astrophysics* by famed Russian scientists. General relativity, properties of matter under astrophysical conditions, stars, and stellar systems. Deep physical insights, clear presentation. 1971 edition. References. 544pp. 5⅝ x 8¼. 0-486-69424-0

Chemistry

THE SCEPTICAL CHYMIST: THE CLASSIC 1661 TEXT, Robert Boyle. Boyle defines the term "element," asserting that all natural phenomena can be explained by the motion and organization of primary particles. 1911 ed. viii+232pp. 5⅜ x 8½.
0-486-42825-7

RADIOACTIVE SUBSTANCES, Marie Curie. Here is the celebrated scientist's doctoral thesis, the prelude to her receipt of the 1903 Nobel Prize. Curie discusses establishing atomic character of radioactivity found in compounds of uranium and thorium; extraction from pitchblende of polonium and radium; isolation of pure radium chloride; determination of atomic weight of radium; plus electric, photographic, luminous, heat, color effects of radioactivity. ii+94pp. 5⅜ x 8½. 0-486-42550-9

CHEMICAL MAGIC, Leonard A. Ford. Second Edition, Revised by E. Winston Grundmeier. Over 100 unusual stunts demonstrating cold fire, dust explosions, much more. Text explains scientific principles and stresses safety precautions. 128pp. 5⅜ x 8½. 0-486-67628-5

THE DEVELOPMENT OF MODERN CHEMISTRY, Aaron J. Ihde. Authoritative history of chemistry from ancient Greek theory to 20th-century innovation. Covers major chemists and their discoveries. 209 illustrations. 14 tables. Bibliographies. Indices. Appendices. 851pp. 5⅜ x 8½. 0-486-64235-6

CATALYSIS IN CHEMISTRY AND ENZYMOLOGY, William P. Jencks. Exceptionally clear coverage of mechanisms for catalysis, forces in aqueous solution, carbonyl- and acyl-group reactions, practical kinetics, more. 864pp. 5⅜ x 8½.
0-486-65460-5

ELEMENTS OF CHEMISTRY, Antoine Lavoisier. Monumental classic by founder of modern chemistry in remarkable reprint of rare 1790 Kerr translation. A must for every student of chemistry or the history of science. 539pp. 5⅜ x 8½. 0-486-64624-6

THE HISTORICAL BACKGROUND OF CHEMISTRY, Henry M. Leicester. Evolution of ideas, not individual biography. Concentrates on formulation of a coherent set of chemical laws. 260pp. 5⅜ x 8½. 0-486-61053-5

A SHORT HISTORY OF CHEMISTRY, J. R. Partington. Classic exposition explores origins of chemistry, alchemy, early medical chemistry, nature of atmosphere, theory of valency, laws and structure of atomic theory, much more. 428pp. 5⅜ x 8½. (Available in U.S. only.) 0-486-65977-1

GENERAL CHEMISTRY, Linus Pauling. Revised 3rd edition of classic first-year text by Nobel laureate. Atomic and molecular structure, quantum mechanics, statistical mechanics, thermodynamics correlated with descriptive chemistry. Problems. 992pp. 5⅜ x 8½. 0-486-65622-5

FROM ALCHEMY TO CHEMISTRY, John Read. Broad, humanistic treatment focuses on great figures of chemistry and ideas that revolutionized the science. 50 illustrations. 240pp. 5⅜ x 8½. 0-486-28690-8

Engineering

DE RE METALLICA, Georgius Agricola. The famous Hoover translation of greatest treatise on technological chemistry, engineering, geology, mining of early modern times (1556). All 289 original woodcuts. 638pp. 6¾ x 11. 0-486-60006-8

FUNDAMENTALS OF ASTRODYNAMICS, Roger Bate et al. Modern approach developed by U.S. Air Force Academy. Designed as a first course. Problems, exercises. Numerous illustrations. 455pp. 5⅜ x 8½. 0-486-60061-0

DYNAMICS OF FLUIDS IN POROUS MEDIA, Jacob Bear. For advanced students of ground water hydrology, soil mechanics and physics, drainage and irrigation engineering and more. 335 illustrations. Exercises, with answers. 784pp. 6⅛ x 9¼. 0-486-65675-6

THEORY OF VISCOELASTICITY (Second Edition), Richard M. Christensen. Complete consistent description of the linear theory of the viscoelastic behavior of materials. Problem-solving techniques discussed. 1982 edition. 29 figures. xiv+364pp. 6⅛ x 9¼. 0-486-42880-X

MECHANICS, J. P. Den Hartog. A classic introductory text or refresher. Hundreds of applications and design problems illuminate fundamentals of trusses, loaded beams and cables, etc. 334 answered problems. 462pp. 5⅜ x 8½. 0-486-60754-2

MECHANICAL VIBRATIONS, J. P. Den Hartog. Classic textbook offers lucid explanations and illustrative models, applying theories of vibrations to a variety of practical industrial engineering problems. Numerous figures. 233 problems, solutions. Appendix. Index. Preface. 436pp. 5⅜ x 8½. 0-486-64785-4

STRENGTH OF MATERIALS, J. P. Den Hartog. Full, clear treatment of basic material (tension, torsion, bending, etc.) plus advanced material on engineering methods, applications. 350 answered problems. 323pp. 5⅜ x 8½. 0-486-60755-0

A HISTORY OF MECHANICS, René Dugas. Monumental study of mechanical principles from antiquity to quantum mechanics. Contributions of ancient Greeks, Galileo, Leonardo, Kepler, Lagrange, many others. 671pp. 5⅜ x 8½. 0-486-65632-2

STABILITY THEORY AND ITS APPLICATIONS TO STRUCTURAL MECHANICS, Clive L. Dym. Self-contained text focuses on Koiter postbuckling analyses, with mathematical notions of stability of motion. Basing minimum energy principles for static stability upon dynamic concepts of stability of motion, it develops asymptotic buckling and postbuckling analyses from potential energy considerations, with applications to columns, plates, and arches. 1974 ed. 208pp. 5⅜ x 8½. 0-486-42541-X

METAL FATIGUE, N. E. Frost, K. J. Marsh, and L. P. Pook. Definitive, clearly written, and well-illustrated volume addresses all aspects of the subject, from the historical development of understanding metal fatigue to vital concepts of the cyclic stress that causes a crack to grow. Includes 7 appendixes. 544pp. 5⅜ x 8½. 0-486-40927-9

Mathematics

FUNCTIONAL ANALYSIS (Second Corrected Edition), George Bachman and Lawrence Narici. Excellent treatment of subject geared toward students with background in linear algebra, advanced calculus, physics and engineering. Text covers introduction to inner-product spaces, normed, metric spaces, and topological spaces; complete orthonormal sets, the Hahn-Banach Theorem and its consequences, and many other related subjects. 1966 ed. 544pp. 6⅛ x 9¼. 0-486-40251-7

ASYMPTOTIC EXPANSIONS OF INTEGRALS, Norman Bleistein & Richard A. Handelsman. Best introduction to important field with applications in a variety of scientific disciplines. New preface. Problems. Diagrams. Tables. Bibliography. Index. 448pp. 5⅜ x 8½. 0-486-65082-0

VECTOR AND TENSOR ANALYSIS WITH APPLICATIONS, A. I. Borisenko and I. E. Tarapov. Concise introduction. Worked-out problems, solutions, exercises. 257pp. 5⅝ x 8¼. 0-486-63833-2

AN INTRODUCTION TO ORDINARY DIFFERENTIAL EQUATIONS, Earl A. Coddington. A thorough and systematic first course in elementary differential equations for undergraduates in mathematics and science, with many exercises and problems (with answers). Index. 304pp. 5⅜ x 8½. 0-486-65942-9

FOURIER SERIES AND ORTHOGONAL FUNCTIONS, Harry F. Davis. An incisive text combining theory and practical example to introduce Fourier series, orthogonal functions and applications of the Fourier method to boundary-value problems. 570 exercises. Answers and notes. 416pp. 5⅜ x 8½. 0-486-65973-9

COMPUTABILITY AND UNSOLVABILITY, Martin Davis. Classic graduate-level introduction to theory of computability, usually referred to as theory of recurrent functions. New preface and appendix. 288pp. 5⅜ x 8½. 0-486-61471-9

ASYMPTOTIC METHODS IN ANALYSIS, N. G. de Bruijn. An inexpensive, comprehensive guide to asymptotic methods–the pioneering work that teaches by explaining worked examples in detail. Index. 224pp. 5⅜ x 8½ 0-486-64221-6

APPLIED COMPLEX VARIABLES, John W. Dettman. Step-by-step coverage of fundamentals of analytic function theory–plus lucid exposition of five important applications: Potential Theory; Ordinary Differential Equations; Fourier Transforms; Laplace Transforms; Asymptotic Expansions. 66 figures. Exercises at chapter ends. 512pp. 5⅜ x 8½. 0-486-64670-X

INTRODUCTION TO LINEAR ALGEBRA AND DIFFERENTIAL EQUATIONS, John W. Dettman. Excellent text covers complex numbers, determinants, orthonormal bases, Laplace transforms, much more. Exercises with solutions. Undergraduate level. 416pp. 5⅜ x 8½. 0-486-65191-6

RIEMANN'S ZETA FUNCTION, H. M. Edwards. Superb, high-level study of landmark 1859 publication entitled "On the Number of Primes Less Than a Given Magnitude" traces developments in mathematical theory that it inspired. xiv+315pp. 5⅜ x 8½. 0-486-41740-9

CATALOG OF DOVER BOOKS

CALCULUS OF VARIATIONS WITH APPLICATIONS, George M. Ewing. Applications-oriented introduction to variational theory develops insight and promotes understanding of specialized books, research papers. Suitable for advanced undergraduate/graduate students as primary, supplementary text. 352pp. 5⅜ x 8½. 0-486-64856-7

COMPLEX VARIABLES, Francis J. Flanigan. Unusual approach, delaying complex algebra till harmonic functions have been analyzed from real variable viewpoint. Includes problems with answers. 364pp. 5⅜ x 8½. 0-486-61388-7

AN INTRODUCTION TO THE CALCULUS OF VARIATIONS, Charles Fox. Graduate-level text covers variations of an integral, isoperimetrical problems, least action, special relativity, approximations, more. References. 279pp. 5⅜ x 8½. 0-486-65499-0

COUNTEREXAMPLES IN ANALYSIS, Bernard R. Gelbaum and John M. H. Olmsted. These counterexamples deal mostly with the part of analysis known as "real variables." The first half covers the real number system, and the second half encompasses higher dimensions. 1962 edition. xxiv+198pp. 5⅜ x 8½. 0-486-42875-3

CATASTROPHE THEORY FOR SCIENTISTS AND ENGINEERS, Robert Gilmore. Advanced-level treatment describes mathematics of theory grounded in the work of Poincaré, R. Thom, other mathematicians. Also important applications to problems in mathematics, physics, chemistry and engineering. 1981 edition. References. 28 tables. 397 black-and-white illustrations. xvii + 666pp. 6⅛ x 9¼. 0-486-67539-4

INTRODUCTION TO DIFFERENCE EQUATIONS, Samuel Goldberg. Exceptionally clear exposition of important discipline with applications to sociology, psychology, economics. Many illustrative examples; over 250 problems. 260pp. 5⅜ x 8½. 0-486-65084-7

NUMERICAL METHODS FOR SCIENTISTS AND ENGINEERS, Richard Hamming. Classic text stresses frequency approach in coverage of algorithms, polynomial approximation, Fourier approximation, exponential approximation, other topics. Revised and enlarged 2nd edition. 721pp. 5⅜ x 8½. 0-486-65241-6

INTRODUCTION TO NUMERICAL ANALYSIS (2nd Edition), F. B. Hildebrand. Classic, fundamental treatment covers computation, approximation, interpolation, numerical differentiation and integration, other topics. 150 new problems. 669pp. 5⅜ x 8½. 0-486-65363-3

THREE PEARLS OF NUMBER THEORY, A. Y. Khinchin. Three compelling puzzles require proof of a basic law governing the world of numbers. Challenges concern van der Waerden's theorem, the Landau-Schnirelmann hypothesis and Mann's theorem, and a solution to Waring's problem. Solutions included. 64pp. 5⅜ x 8½. 0-486-40026-3

THE PHILOSOPHY OF MATHEMATICS: AN INTRODUCTORY ESSAY, Stephan Körner. Surveys the views of Plato, Aristotle, Leibniz & Kant concerning propositions and theories of applied and pure mathematics. Introduction. Two appendices. Index. 198pp. 5⅜ x 8½. 0-486-25048-2

TENSOR CALCULUS, J.L. Synge and A. Schild. Widely used introductory text covers spaces and tensors, basic operations in Riemannian space, non-Riemannian spaces, etc. 324pp. 5⅝ x 8¼. 0-486-63612-7

ORDINARY DIFFERENTIAL EQUATIONS, Morris Tenenbaum and Harry Pollard. Exhaustive survey of ordinary differential equations for undergraduates in mathematics, engineering, science. Thorough analysis of theorems. Diagrams. Bibliography. Index. 818pp. 5⅜ x 8½. 0-486-64940-7

INTEGRAL EQUATIONS, F. G. Tricomi. Authoritative, well-written treatment of extremely useful mathematical tool with wide applications. Volterra Equations, Fredholm Equations, much more. Advanced undergraduate to graduate level. Exercises. Bibliography. 238pp. 5⅜ x 8½. 0-486-64828-1

FOURIER SERIES, Georgi P. Tolstov. Translated by Richard A. Silverman. A valuable addition to the literature on the subject, moving clearly from subject to subject and theorem to theorem. 107 problems, answers. 336pp. 5⅜ x 8½. 0-486-63317-9

INTRODUCTION TO MATHEMATICAL THINKING, Friedrich Waismann. Examinations of arithmetic, geometry, and theory of integers; rational and natural numbers; complete induction; limit and point of accumulation; remarkable curves; complex and hypercomplex numbers, more. 1959 ed. 27 figures. xii+260pp. 5⅜ x 8½. 0-486-63317-9

POPULAR LECTURES ON MATHEMATICAL LOGIC, Hao Wang. Noted logician's lucid treatment of historical developments, set theory, model theory, recursion theory and constructivism, proof theory, more. 3 appendixes. Bibliography. 1981 edition. ix + 283pp. 5⅜ x 8½. 0-486-67632-3

CALCULUS OF VARIATIONS, Robert Weinstock. Basic introduction covering isoperimetric problems, theory of elasticity, quantum mechanics, electrostatics, etc. Exercises throughout. 326pp. 5⅜ x 8½. 0-486-63069-2

THE CONTINUUM: A CRITICAL EXAMINATION OF THE FOUNDATION OF ANALYSIS, Hermann Weyl. Classic of 20th-century foundational research deals with the conceptual problem posed by the continuum. 156pp. 5⅜ x 8½. 0-486-67982-9

CHALLENGING MATHEMATICAL PROBLEMS WITH ELEMENTARY SOLUTIONS, A. M. Yaglom and I. M. Yaglom. Over 170 challenging problems on probability theory, combinatorial analysis, points and lines, topology, convex polygons, many other topics. Solutions. Total of 445pp. 5⅜ x 8½. Two-vol. set. Vol. I: 0-486-65536-9 Vol. II: 0-486-65537-7

INTRODUCTION TO PARTIAL DIFFERENTIAL EQUATIONS WITH APPLICATIONS, E. C. Zachmanoglou and Dale W. Thoe. Essentials of partial differential equations applied to common problems in engineering and the physical sciences. Problems and answers. 416pp. 5⅜ x 8½. 0-486-65251-3

THE THEORY OF GROUPS, Hans J. Zassenhaus. Well-written graduate-level text acquaints reader with group-theoretic methods and demonstrates their usefulness in mathematics. Axioms, the calculus of complexes, homomorphic mapping, p-group theory, more. 276pp. 5⅜ x 8½. 0-486-40922-8

Math–Decision Theory, Statistics, Probability

ELEMENTARY DECISION THEORY, Herman Chernoff and Lincoln E. Moses. Clear introduction to statistics and statistical theory covers data processing, probability and random variables, testing hypotheses, much more. Exercises. 364pp. 5⅜ x 8½. 0-486-65218-1

STATISTICS MANUAL, Edwin L. Crow et al. Comprehensive, practical collection of classical and modern methods prepared by U.S. Naval Ordnance Test Station. Stress on use. Basics of statistics assumed. 288pp. 5⅜ x 8½. 0-486-60599-X

SOME THEORY OF SAMPLING, William Edwards Deming. Analysis of the problems, theory and design of sampling techniques for social scientists, industrial managers and others who find statistics important at work. 61 tables. 90 figures. xvii +602pp. 5⅜ x 8½. 0-486-64684-X

LINEAR PROGRAMMING AND ECONOMIC ANALYSIS, Robert Dorfman, Paul A. Samuelson and Robert M. Solow. First comprehensive treatment of linear programming in standard economic analysis. Game theory, modern welfare economics, Leontief input-output, more. 525pp. 5⅜ x 8½. 0-486-65491-5

PROBABILITY: AN INTRODUCTION, Samuel Goldberg. Excellent basic text covers set theory, probability theory for finite sample spaces, binomial theorem, much more. 360 problems. Bibliographies. 322pp. 5⅜ x 8½. 0-486-65252-1

GAMES AND DECISIONS: INTRODUCTION AND CRITICAL SURVEY, R. Duncan Luce and Howard Raiffa. Superb nontechnical introduction to game theory, primarily applied to social sciences. Utility theory, zero-sum games, n-person games, decision-making, much more. Bibliography. 509pp. 5⅜ x 8½. 0-486-65943-7

INTRODUCTION TO THE THEORY OF GAMES, J. C. C. McKinsey. This comprehensive overview of the mathematical theory of games illustrates applications to situations involving conflicts of interest, including economic, social, political, and military contexts. Appropriate for advanced undergraduate and graduate courses; advanced calculus a prerequisite. 1952 ed. x+372pp. 5⅜ x 8½. 0-486-42811-7

FIFTY CHALLENGING PROBLEMS IN PROBABILITY WITH SOLUTIONS, Frederick Mosteller. Remarkable puzzlers, graded in difficulty, illustrate elementary and advanced aspects of probability. Detailed solutions. 88pp. 5⅜ x 8½. 65355-2

PROBABILITY THEORY: A CONCISE COURSE, Y. A. Rozanov. Highly readable, self-contained introduction covers combination of events, dependent events, Bernoulli trials, etc. 148pp. 5⅜ x 8¼. 0-486-63544-9

STATISTICAL METHOD FROM THE VIEWPOINT OF QUALITY CONTROL, Walter A. Shewhart. Important text explains regulation of variables, uses of statistical control to achieve quality control in industry, agriculture, other areas. 192pp. 5⅜ x 8½. 0-486-65232-7

Math–Geometry and Topology

ELEMENTARY CONCEPTS OF TOPOLOGY, Paul Alexandroff. Elegant, intuitive approach to topology from set-theoretic topology to Betti groups; how concepts of topology are useful in math and physics. 25 figures. 57pp. 5⅜ x 8½. 0-486-60747-X

COMBINATORIAL TOPOLOGY, P. S. Alexandrov. Clearly written, well-organized, three-part text begins by dealing with certain classic problems without using the formal techniques of homology theory and advances to the central concept, the Betti groups. Numerous detailed examples. 654pp. 5⅜ x 8½. 0-486-40179-0

EXPERIMENTS IN TOPOLOGY, Stephen Barr. Classic, lively explanation of one of the byways of mathematics. Klein bottles, Moebius strips, projective planes, map coloring, problem of the Koenigsberg bridges, much more, described with clarity and wit. 43 figures. 210pp. 5⅜ x 8½. 0-486-25933-1

THE GEOMETRY OF RENÉ DESCARTES, René Descartes. The great work founded analytical geometry. Original French text, Descartes's own diagrams, together with definitive Smith-Latham translation. 244pp. 5⅜ x 8½. 0-486-60068-8

EUCLIDEAN GEOMETRY AND TRANSFORMATIONS, Clayton W. Dodge. This introduction to Euclidean geometry emphasizes transformations, particularly isometries and similarities. Suitable for undergraduate courses, it includes numerous examples, many with detailed answers. 1972 ed. viii+296pp. 6⅛ x 9¼. 0-486-43476-1

PRACTICAL CONIC SECTIONS: THE GEOMETRIC PROPERTIES OF ELLIPSES, PARABOLAS AND HYPERBOLAS, J. W. Downs. This text shows how to create ellipses, parabolas, and hyperbolas. It also presents historical background on their ancient origins and describes the reflective properties and roles of curves in design applications. 1993 ed. 98 figures. xii+100pp. 6½ x 9¼. 0-486-42876-1

THE THIRTEEN BOOKS OF EUCLID'S ELEMENTS, translated with introduction and commentary by Sir Thomas L. Heath. Definitive edition. Textual and linguistic notes, mathematical analysis. 2,500 years of critical commentary. Unabridged. 1,414pp. 5⅜ x 8½. Three-vol. set.
Vol. I: 0-486-60088-2 Vol. II: 0-486-60089-0 Vol. III: 0-486-60090-4

SPACE AND GEOMETRY: IN THE LIGHT OF PHYSIOLOGICAL, PSYCHOLOGICAL AND PHYSICAL INQUIRY, Ernst Mach. Three essays by an eminent philosopher and scientist explore the nature, origin, and development of our concepts of space, with a distinctness and precision suitable for undergraduate students and other readers. 1906 ed. vi+148pp. 5⅜ x 8½. 0-486-43909-7

GEOMETRY OF COMPLEX NUMBERS, Hans Schwerdtfeger. Illuminating, widely praised book on analytic geometry of circles, the Moebius transformation, and two-dimensional non-Euclidean geometries. 200pp. 5⅝ x 8¼. 0-486-63830-8

DIFFERENTIAL GEOMETRY, Heinrich W. Guggenheimer. Local differential geometry as an application of advanced calculus and linear algebra. Curvature, transformation groups, surfaces, more. Exercises. 62 figures. 378pp. 5⅜ x 8½. 0-486-63433-7

History of Math

THE WORKS OF ARCHIMEDES, Archimedes (T. L. Heath, ed.). Topics include the famous problems of the ratio of the areas of a cylinder and an inscribed sphere; the measurement of a circle; the properties of conoids, spheroids, and spirals; and the quadrature of the parabola. Informative introduction. clxxxvi+326pp. 5⅜ x 8½. 0-486-42084-1

A SHORT ACCOUNT OF THE HISTORY OF MATHEMATICS, W. W. Rouse Ball. One of clearest, most authoritative surveys from the Egyptians and Phoenicians through 19th-century figures such as Grassman, Galois, Riemann. Fourth edition. 522pp. 5⅜ x 8½. 0-486-20630-0

THE HISTORY OF THE CALCULUS AND ITS CONCEPTUAL DEVELOPMENT, Carl B. Boyer. Origins in antiquity, medieval contributions, work of Newton, Leibniz, rigorous formulation. Treatment is verbal. 346pp. 5⅜ x 8½. 0-486-60509-4

THE HISTORICAL ROOTS OF ELEMENTARY MATHEMATICS, Lucas N. H. Bunt, Phillip S. Jones, and Jack D. Bedient. Fundamental underpinnings of modern arithmetic, algebra, geometry and number systems derived from ancient civilizations. 320pp. 5⅜ x 8½. 0-486-25563-8

A HISTORY OF MATHEMATICAL NOTATIONS, Florian Cajori. This classic study notes the first appearance of a mathematical symbol and its origin, the competition it encountered, its spread among writers in different countries, its rise to popularity, its eventual decline or ultimate survival. Original 1929 two-volume edition presented here in one volume. xxviii+820pp. 5⅜ x 8½. 0-486-67766-4

GAMES, GODS & GAMBLING: A HISTORY OF PROBABILITY AND STATISTICAL IDEAS, F. N. David. Episodes from the lives of Galileo, Fermat, Pascal, and others illustrate this fascinating account of the roots of mathematics. Features thought-provoking references to classics, archaeology, biography, poetry. 1962 edition. 304pp. 5⅜ x 8½. (Available in U.S. only.) 0-486-40023-9

OF MEN AND NUMBERS: THE STORY OF THE GREAT MATHEMATICIANS, Jane Muir. Fascinating accounts of the lives and accomplishments of history's greatest mathematical minds–Pythagoras, Descartes, Euler, Pascal, Cantor, many more. Anecdotal, illuminating. 30 diagrams. Bibliography. 256pp. 5⅜ x 8½. 0-486-28973-7

HISTORY OF MATHEMATICS, David E. Smith. Nontechnical survey from ancient Greece and Orient to late 19th century; evolution of arithmetic, geometry, trigonometry, calculating devices, algebra, the calculus. 362 illustrations. 1,355pp. 5⅜ x 8½. Two-vol. set. Vol. I: 0-486-20429-4 Vol. II: 0-486-20430-8

A CONCISE HISTORY OF MATHEMATICS, Dirk J. Struik. The best brief history of mathematics. Stresses origins and covers every major figure from ancient Near East to 19th century. 41 illustrations. 195pp. 5⅜ x 8½. 0-486-60255-9

Physics

OPTICAL RESONANCE AND TWO-LEVEL ATOMS, L. Allen and J. H. Eberly. Clear, comprehensive introduction to basic principles behind all quantum optical resonance phenomena. 53 illustrations. Preface. Index. 256pp. 5⅜ x 8½. 0-486-65533-4

QUANTUM THEORY, David Bohm. This advanced undergraduate-level text presents the quantum theory in terms of qualitative and imaginative concepts, followed by specific applications worked out in mathematical detail. Preface. Index. 655pp. 5⅜ x 8½. 0-486-65969-0

ATOMIC PHYSICS (8th EDITION), Max Born. Nobel laureate's lucid treatment of kinetic theory of gases, elementary particles, nuclear atom, wave-corpuscles, atomic structure and spectral lines, much more. Over 40 appendices, bibliography. 495pp. 5⅜ x 8½. 0-486-65984-4

A SOPHISTICATE'S PRIMER OF RELATIVITY, P. W. Bridgman. Geared toward readers already acquainted with special relativity, this book transcends the view of theory as a working tool to answer natural questions: What is a frame of reference? What is a "law of nature"? What is the role of the "observer"? Extensive treatment, written in terms accessible to those without a scientific background. 1983 ed. xlviii+172pp. 5⅜ x 8½. 0-486-42549-5

AN INTRODUCTION TO HAMILTONIAN OPTICS, H. A. Buchdahl. Detailed account of the Hamiltonian treatment of aberration theory in geometrical optics. Many classes of optical systems defined in terms of the symmetries they possess. Problems with detailed solutions. 1970 edition. xv + 360pp. 5⅜ x 8½. 0-486-67597-1

PRIMER OF QUANTUM MECHANICS, Marvin Chester. Introductory text examines the classical quantum bead on a track: its state and representations; operator eigenvalues; harmonic oscillator and bound bead in a symmetric force field; and bead in a spherical shell. Other topics include spin, matrices, and the structure of quantum mechanics; the simplest atom; indistinguishable particles; and stationary-state perturbation theory. 1992 ed. xiv+314pp. 6⅛ x 9¼. 0-486-42878-8

LECTURES ON QUANTUM MECHANICS, Paul A. M. Dirac. Four concise, brilliant lectures on mathematical methods in quantum mechanics from Nobel Prize-winning quantum pioneer build on idea of visualizing quantum theory through the use of classical mechanics. 96pp. 5⅜ x 8½. 0-486-41713-1

THIRTY YEARS THAT SHOOK PHYSICS: THE STORY OF QUANTUM THEORY, George Gamow. Lucid, accessible introduction to influential theory of energy and matter. Careful explanations of Dirac's anti-particles, Bohr's model of the atom, much more. 12 plates. Numerous drawings. 240pp. 5⅜ x 8½. 0-486-24895-X

ELECTRONIC STRUCTURE AND THE PROPERTIES OF SOLIDS: THE PHYSICS OF THE CHEMICAL BOND, Walter A. Harrison. Innovative text offers basic understanding of the electronic structure of covalent and ionic solids, simple metals, transition metals and their compounds. Problems. 1980 edition. 582pp. 6⅛ x 9¼. 0-486-66021-4

CATALOG OF DOVER BOOKS

A TREATISE ON ELECTRICITY AND MAGNETISM, James Clerk Maxwell. Important foundation work of modern physics. Brings to final form Maxwell's theory of electromagnetism and rigorously derives his general equations of field theory. 1,084pp. 5⅜ x 8½. Two-vol. set. Vol. I: 0-486-60636-8 Vol. II: 0-486-60637-6

QUANTUM MECHANICS: PRINCIPLES AND FORMALISM, Roy McWeeny. Graduate student-oriented volume develops subject as fundamental discipline, opening with review of origins of Schrödinger's equations and vector spaces. Focusing on main principles of quantum mechanics and their immediate consequences, it concludes with final generalizations covering alternative "languages" or representations. 1972 ed. 15 figures. xi+155pp. 5⅜ x 8½. 0-486-42829-X

INTRODUCTION TO QUANTUM MECHANICS With Applications to Chemistry, Linus Pauling & E. Bright Wilson, Jr. Classic undergraduate text by Nobel Prize winner applies quantum mechanics to chemical and physical problems. Numerous tables and figures enhance the text. Chapter bibliographies. Appendices. Index. 468pp. 5⅜ x 8½. 0-486-64871-0

METHODS OF THERMODYNAMICS, Howard Reiss. Outstanding text focuses on physical technique of thermodynamics, typical problem areas of understanding, and significance and use of thermodynamic potential. 1965 edition. 238pp. 5⅜ x 8½. 0-486-69445-3

THE ELECTROMAGNETIC FIELD, Albert Shadowitz. Comprehensive undergraduate text covers basics of electric and magnetic fields, builds up to electromagnetic theory. Also related topics, including relativity. Over 900 problems. 768pp. 5⅜ x 8¼. 0-486-65660-8

GREAT EXPERIMENTS IN PHYSICS: FIRSTHAND ACCOUNTS FROM GALILEO TO EINSTEIN, Morris H. Shamos (ed.). 25 crucial discoveries: Newton's laws of motion, Chadwick's study of the neutron, Hertz on electromagnetic waves, more. Original accounts clearly annotated. 370pp. 5⅜ x 8½. 0-486-25346-5

EINSTEIN'S LEGACY, Julian Schwinger. A Nobel Laureate relates fascinating story of Einstein and development of relativity theory in well-illustrated, nontechnical volume. Subjects include meaning of time, paradoxes of space travel, gravity and its effect on light, non-Euclidean geometry and curving of space-time, impact of radio astronomy and space-age discoveries, and more. 189 b/w illustrations. xiv+250pp. 8⅜ x 9¼. 0-486-41974-6

STATISTICAL PHYSICS, Gregory H. Wannier. Classic text combines thermodynamics, statistical mechanics and kinetic theory in one unified presentation of thermal physics. Problems with solutions. Bibliography. 532pp. 5⅜ x 8½. 0-486-65401-X

Paperbound unless otherwise indicated. Available at your book dealer, online at **www.doverpublications.com**, or by writing to Dept. GI, Dover Publications, Inc., 31 East 2nd Street, Mineola, NY 11501. For current price information or for free catalogues (please indicate field of interest), write to Dover Publications or log on to **www.doverpublications.com** and see every Dover book in print. Dover publishes more than 500 books each year on science, elementary and advanced mathematics, biology, music, art, literary history, social sciences, and other areas.